Fish Swimming in Turbulent Waters

Fish Swimming in Turbulent Waters

Hydraulic Engineering Guidelines to Assist Upstream Passage of Small-Bodied Fish Species in Standard Box Culverts

Hubert Chanson and Xinqian Leng

The University of Queensland, School of Civil Engineering, Brisbane, Australia

"Fish awaiting passage" by Harry Chia Huang

CRC Press
Taylor & Francis Group
Boca Raton London New York Leiden

CRC Press is an imprint of the
Taylor & Francis Group, an **informa** business

A BALKEMA BOOK

CRC Press/Balkema is an imprint of the Taylor & Francis Group, an informa business

© 2021 Taylor & Francis Group, London, UK

Typeset by Apex CoVantage, LLC

Library of Congress Cataloging-in-Publication Data
Applied for

Published by: CRC Press/Balkema
                Schipholweg 107C, 2316 XC Leiden, The Netherlands
                e-mail: Pub.NL@taylorandfrancis.com
                www.routledge.com – www.taylorandfrancis.com

ISBN: 978-0-367-46573-5 (hbk)
ISBN: 978-0-367-54606-9 (pbk)
ISBN: 978-1-003-02969-4 (ebk)
DOI: 10.1201/9781003029694
DOI: https://doi.org/10.1201/9781003029694

We would like to dedicate the book to:

Hubert's family

À mes parents
To Ya Hui.
Pour Bernard, Nicole et André.

Xinqian's family and friends

My parents; my mother Fan Liu and father Chuanjun Leng
My grandparents
My friends; reasons I enjoy life

Contents

Symbols

The following symbols are used in this book:

A	• channel cross-section area (m^2)
	• relative low-velocity zone area: $A = LVZ/(B_{cell} \times d_{barrel})$
a	acceleration (m/s^2); $a = \partial U/\partial t$
B	channel width (m)
B_{cell}	internal width (m) of culvert barrel cell
B_{min}	minimum total internal barrel width (m)
C	color function
$C_{Chézy}$	Chézy friction coefficient ($m^{1/2}.s$)
d	water depth (m)
d_{barrel}	water depth (m) in the culvert barrel
d_c	critical flow depth (m); for a rectangular channel:

$$d_c = \sqrt[3]{\frac{q^2}{g}}$$

D_{cell}	internal height (m) of culvert barrel cell
D_H	hydraulic diameter (m): $D_H = 4 \times A/P_w$
d_{tw}	tailwater water depth (m)
E	specific energy (m): $E = H - z_o$
f	Darcy-Weisbach friction factor
g	gravity acceleration (m/s^2); in Brisbane, $g = 9.794$ m/s^2
H	total head (m)
h	afflux (m)
h_b	triangular corner baffle height (m)
h_f	vertical height (m) of the targeted fish species
h_{max}	maximum acceptable afflux (m)
k_e	entrance loss coefficient
K_{out}	exit loss coefficient
k_s	equivalent sand roughness height (m)
L	length (m)
L_b	longitudinal spacing (m) between baffles
L_{barrel}	barrel length (m)
L_{culv}	entire culvert length (m)
L_f	total fish length (m)

LVZ	low-velocity zone cross-section area (m²), where $0 < V_x < U_{fish}$, in the culvert barrel cell
m_f	fish mass (kg)
N	• number of dimensional variables • inverse of velocity power law
N_{cell}	number of (identical) culvert barrel cells
$(N_{cell})_{des}$	number of culvert barrel cells for optimum hydraulic engineering design (i.e. flood capacity only)
n_{GM}	Gauckler-Manning coefficient (s/m$^{1/3}$)
P_w	wetted perimeter (m)
Q	water discharge (m³/s)
q	water discharge per unit width (m²/s), or unit discharge: q = Q/B
Q_{cell}	water discharge (m³/s) per barrel cell
Q_{des}	design discharge (m³/s)
Q_{min}	minimum discharge (m³/s) for fish passage
Q_T	threshold discharge (m³/s) for fish passage, only considered for $Q < Q_T < Q_{des}$
Q_{30}	flow rate (m³/s) occurring no more than 30 days per year
Q_{330}	flow rate (m³/s) occurring no more than 330 days per year
S_f	friction slope
S_o	bed slope: $S_o = \sin\theta$
T	water temperature (Celsius)
$T_{50\%}$	storm event period (s) during which the relative water level is larger than 50% of the maximum water elevation relative to the base flow level
U	fish speed (m/s)
u	fish speed fluctuations (m/s)
U_{fish}	characteristic fish swimming speed (m/s), set by fishery agency based upon fish swimming performance data
$(U_{fish})_{min}$	minimum characteristic fish speed (m/s) of a fish guild
V	flow velocity (m/s)
v	water velocity fluctuation (m/s)
V_{mean}	cross-sectional averaged velocity (m/s), also called bulk velocity
$(V_{mean})_{des}$	design bulk velocity (m/s) in culvert barrel
V_x	local longitudinal velocity component (m/s) positive downstream
x	longitudinal distance (m) positive downstream
y	transverse distance (m) positive toward the left sidewall
z	vertical elevation (m) positive upwards
z_o	bed elevation (m)
ΔH	total head loss (m)
κ	von Karman constant: $\kappa = 0.4$
μ	water dynamic viscosity (Pa.s)
θ	angle between bed and horizontal
ρ	water density (kg/m³)

Subscript

barrel	barrel flow conditions
cell	barrel cell characteristics

des   design flow conditions
entry   barrel entry flow conditions
exit   barrel exit flow conditions
f   fish property
hw   headwater conditions (i.e. upstream flood plain conditions)
in   inlet downstream end's flow conditions
tw   tailwater conditions (i.e. downstream flood plain conditions)
w   water
x   longitudinal direction

Superscript

ic   inlet control conditions
oc   outlet control conditions

Abbreviations

AEP   annual exceedance probability
AM   annual maximum
AMS   annual maxima series
ARI   average recurrence interval
ARR   Australian rainfall and runoff
CFD   computational fluid dynamics
EY   number of exceedances per year
LPVZ   low-positive-velocity zone
LVZ   low-velocity zone
min   minute
NVZ   negative-velocity zone
POT   peak-over-threshold
s   second
THL   total head line
TWL   tailwater level
WWII   World War Two
1D   one-dimensional
2D   two-dimensional
3D   three-dimensional

d/dt	deltaic flow conditions
entr	partial entry flow conditions
exit	partial exit flow conditions
C	fish property
bc	clearance conditions (i.e. upstream flood plain conditions)
in	inlet downstream end's flow conditions
tw	tailwater conditions (i.e. downstream flood plain conditions)
wter	water
x	longitudinal direction

Superscript

ic	inlet control conditions
oc	outlet control conditions

Abbreviations

AEP	annual exceedance probability
AM	annual maximum
AMS	annual maximum series
ARI	average recurrence interval
ARR	Australian rainfall and runoff
CFD	computational fluid dynamics
E?	number of exceedances per year
LPVZ	low-positive-velocity zone
LVZ	low-velocity zone
min	minute
NVZ	negative-velocity zone
POT	peak over-threshold
s	second
THL	total head line
TWL	tailwater level
WWII	World War Two
1D	one-dimensional
2D	two-dimensional
3D	three-dimensional

Acknowledgments

Both authors thank Professor Colin J. Apelt (The University of Queensland, Australia), Professor Blake Tullis (Utah State University, USA), Robert Janssen (Bechtel Australia), and Dr. Chris Katopodis (Katopodis Ecohydraulics, Canada) for the valuable exchanges and comments. They further acknowledge the helpful discussions with Dr. Matthew Gordos and Marcus Riches (NSW DPI Fisheries), including detailed comments, as well as fruitful exchanges with Emeritus Professor Colin J. Apelt (The University of Queensland).

Further comments from discussions with a number of professional engineers were considered.

HC thanks the many student volunteers, research students, and early-career researchers whose respective contributions made this project possible:

Civil Engineering students: Angela Arum, Laura Beckingham,[1,2] Joseph Cabonce, Michael Cheung, Joseph Dowling, Ramith Fernando, Ruben Freire, Caitlyn Johnson, Urvisha Kiri,[3] Matilda Meppem, Thi My Tram Ngo, Carlos Sailema, Pedro Sanchez, Rui Shi,[4,5] Jui Jie Tan, Amelia Tatham, Jee Sam Tiew, Warren Uys, Johann Von Brandis-Martini, Eric Wu, Tianwei Yin.[6]

Civil Engineering research fellows: Dr. Xinqian (Sophia) Leng[7], Dr. Hang Wang,[8,9] Dr. Gangfu Zhang.

Commentaries were sought from engineers, biologists, ecologists, environmentalists, and regulators working in private and public sectors. Contributors were asked to comment on the most relevant key message(s) as well as some important key points. Both authors thank the following individuals for their input:

Emeritus Professor Colin J. Apelt (The University of Queensland, Australia), Dr. Cindy Baker (NIWA, New Zealand), Professor Daniel Bung (FH Aachen University of Applied Sciences, Germany), Dr. Hang Wang (Sichuan University, China).

The following institutions and individuals are thanked for providing data, illustrations, and photographs of interest:

Concrete Pipe Association of Australasia, Harry Chia Huang ("Fish awaiting passage"), NSW DPI Fisheries, Water NSW.

The financial support of the Australian Research Council (Grant LP140100225), NSW DPI Fisheries, NSW Road and Maritime Safety, QLD Transport and Main Roads (Grant TMTHF1805), and the School of Civil Engineering at the University of Queensland is acknowledged.

Notes

1 2018 UQ Civil Engineering Honour Board for Outstanding Results in Fluids.
2 2018 UQ Civil Engineering Honour Board for Outstanding Results in Environmental.
3 Engineers Australia's National Committee on Water Engineering 2017 Scholarship Award.
4 Engineers Australia's 2017 W.H.R. Nimmo Award.
5 2017 Sir Charles Burton Prize, The University of Queensland.
6 2018 UQ Civil Engineering Honour Board for Outstanding Student (2nd Year).
7 2018 Institution of Civil Engineers (UK) Baker Medal.
8 2018 Institution of Civil Engineers (UK) Baker Medal.
9 2014 Lorenz G. Straub Award (USA).

Declarations of interest

In line with the recommendations of the Office of the Commonwealth Ombudsman (Australia), Hubert Chanson declares a competing interest and conflict of interest with Craig E. Franklin.

About the authors

Hubert Chanson is Professor of Civil Engineering at the University of Queensland, where he has been since 1990, having previously enjoyed an industrial career for six years. His main field of expertise is environmental fluid mechanics and hydraulic engineering, both in terms of theoretical fundamentals and physical and numerical modeling. He leads a group of 5 to 10 researchers, largely targeting flows around hydraulic structures, two-phase (gas–liquid and solid–liquid) free-surface flows, and turbulence in steady and unsteady open channel flows, using computation, lab-scale experiments, field work,  and analysis. He has published over 1200 peer-reviewed publications, including more than 20 books. He serves on the editorial boards of *International Journal of Multiphase Flow*, *Flow Measurement and Instrumentation*, and *Environmental Fluid Mechanics*, the latter of which he is currently a senior editor. www.uq.edu.au/~c2hchans

Dr. Xinqian (Sophia) Leng is a postdoctoral research fellow at the University of Queensland. Her research interests include experimental investigations of unsteady rapidly varied open channel flows, computational fluid dynamics (CFD) modeling of bores and hydraulic structures, and field investigations of tidal bores. She authored 60 peer-reviewed papers, including 23 international scientific journal articles. Dr. Leng is the recipient of the 2018 Baker Medal, Institution of Civil Engineers, UK, for the paper in the journal *Engineering and Computational Mechanics*, Proceedings of the Institution of Civil Engineers, UK. She was awarded a 2019–2021 Idex international research fellowship from the University of Bordeaux (France).  Xinqian is actively involved in international collaborations with overseas research institutions such as the University of Bordeaux and Zhejiang Institution of Hydraulics and Estuary (China), as well as interdisciplinary cooperation with industrial partners.

Foreword

Emeritus Professor Colin J. Apelt

This monograph *Fish Swimming in Turbulent Waters: Hydraulic Engineering Guidelines to assist Upstream Passage of Small-Bodied Fish Species in Standard Box Culverts* includes all of the information required to complete the design of fish-friendly standard box culverts. It combines the details of standard hydraulic engineering design of box culverts with the details of their design for the upstream passage of small-body-mass fish. It is complete and self-contained – there is no need to refer to other sources. However, as stressed by the authors, the design of a culvert that is intended to be constructed would require the certification of a professional civil engineer.

The great virtue of the methodology detailed in the monograph is that it uses standard box culverts without resorting to expensive modifications. (Alternative approaches that incorporate various forms of corner baffles or artificial roughening of bottom and sidewall boundaries are summarized in Appendix E.)

The introduction provides an excellent overview of the issues involved, of their importance, and of their treatment in the chapters that follow. The conflicting requirements for the standard hydraulic design of box culverts for roadway crossings of waterways versus those for the upstream passage of small-body-mass fish are described – standard hydraulic engineering design seeks to minimize the size of the culvert, but this usually creates barriers to the upstream passage of fish. The negative impacts these have on the access of fish to feeding and breeding habitats and on the conservation of threatened species are described. The guidelines for freshwater fish passage in standard culverts are provided for a number of fish species from France and Germany, five states in the USA, and Australia, showing that this is a worldwide issue and that the approach given in this monograph is of worldwide relevance.

Chapter 2 details the standard approach to the hydraulic design of box culverts. This material would be familiar to hydraulic design engineers, but its inclusion makes the whole procedure understandable to readers who do not have this background.

Chapter 3 provides a comprehensive review of laboratory studies of small-body-mass fish swimming in conditions encountered in box culverts. Details of many of these studies are given in Appendix C.

The key contribution of the monograph, its core, is in Chapter 4. It details a new approach that modifies the standard culvert design to be fish friendly. In summary, the number of cells is increased beyond those determined by the standard design until conditions are achieved that allow upstream passage of fish at a flow that is less than the design flow. The essence of the approach is to achieve an acceptable proportion of the flow area with velocities less than that considered the limit at which the target fish can achieve upstream passage and with dimensions large enough for the fish to be able to swim. The specification of these conditions, including the choice of the less-than-design flow, is ultimately the responsibility of authorities concerned with fisheries. Details of the velocity distribution and of the required dimensions could be determined either by physical or computational fluid dynamics (CFD) modeling, but this is not practicable for normal hydraulic engineering design. To achieve a practicable approach, the authors have collected the data from physical and CFD modeling and summarized this in a diagram that gives the percentage of flow area for which the velocity is less than a percentage of the bulk mean velocity. This has been consolidated into a monotonic relationship considered satisfactory for design purposes. The key dimension used in the design is the length of the corner bisector of the low-velocity region. This also has been determined from all the available results from physical and CFD modeling.

The step-by-step example of the design procedure described in Chapter 4 is provided with application to a specific example in Chapter 5. In this example, the flow chosen as that for which flow conditions must be suitable for upstream fish passage is 10% of the design flow, and the proportion of flow area in which these conditions exist is chosen to be not less than 15% of the flow area. These percentages are for the example; those to be used in a particular application would require input from experts on fisheries. In this example the number of culvert cells required by the standard design is 9, whereas 12 cells are required to satisfy the specified requirements for upstream fish passage.

Some further aspects of the hydraulic calculations of less-than-design flows in box culverts are discussed in Appendix B.

While not necessary for the understanding of the design procedures in Chapters 4 and 5, Appendix D and Appendix F provide greater insight into the basis for the design procedure.

This monograph is an excellent example of the application of hydraulic engineering principles to the solution of an important environmental issue. It provides a practical methodology for the design of box culverts that will allow the upstream passage of small-body-mass fish. It is essential reading for all who are concerned with the preservation of native fish species.

Colin J. Apelt
Emeritus Professor of Civil Engineering,
The University of Queensland

Chapter 1

Introduction

1.1 Presentation

Low-level river crossings are important for delivering a range of important socioeconomic services, including transportation and hydrological control. These structures are also known to have negative impacts on freshwater river system morphology and ecology, including the blockage of upstream fish passage (Warren and Pardew, 1998, Anderson *et al.*, 2012). The manner in which waterway crossings block fish movement include perched outlet,[1] high velocity and insufficient water depth in the culvert barrel, debris accumulation at the culvert inlet, and standing waves in the outlet or inlet (Behlke *et al.*, 1991; Olsen and Tullis, 2013). And it is closely linked to the targeted fish species. For small, weak-swimming fish species, the upstream traversability of the culvert barrel is too often a major obstacle, especially because of the high water velocities. In order to restore upstream fish passage in culverts over the widest extent possible, the thrust of this monograph is to apply a physically based design methodology to yield cost-effective culvert structures, with the aim to maintain and restore waterway connectivity for a range of small-bodied and juvenile native fish species.

A culvert is a covered channel designed to pass flood waters, drainage flows, and natural streams through embankment structures (e.g. roadway, railroad) (Fig. 1.1). (Appendix A details a glossary of technical terms.) The cross-sectional shape of the culvert barrel may be circular (pipe) or rectangular (box and multicell box), and may be designed as a single-cell or multiple-cell structure. In terms of hydraulic engineering, a box culvert is basically a covered rectangular channel, with a converging section at the entrance, called the inlet, and a diverging section at the exit, called the outlet. The culvert channel is typically narrower than the natural river channel. The narrowest part of the culvert is the barrel or throat. Sometimes, rectangular cells are placed side by side to increase the discharge capacity (i.e. a multicell box culvert). Figures 1.1A and 1.1B present typical examples of modern multicell box culverts, and further examples are illustrated in Chanson and Leng (2019). Figure 1.2 shows some culvert operations for medium to large discharges. In one case (Fig. 1.2A), the road embankment was overtopped, and the discharge at the time of the photograph was larger than the design discharge.[2] The movie CIMG50.48.mov (Appendix F) illustrates a box culvert operation at full capacity.

During operation, the fluid flow motion in a culvert is complicated because of the boundary conditions and flow turbulence. The prediction of the fluid dynamics is challenging due to the broad range of culvert shapes and designs (Figs. 1.1 and 1.2). For discharges up to the design discharge, the culvert structure should operate as a free-surface flow. The open channel fluid dynamics are intricate because of the complicated interactions between the water and a number of mechanisms, including the boundary friction, gravity, and turbulence (Rouse, 1938; Chow, 1959; Henderson, 1966). Traditionally, open channel flows have been modeled based upon one-dimensional, depth-averaged equations, which predict the mean

(A) Culvert inlet below Kate Street, Indooroopilly QLD, Australia, on 15 October 2018

(B) Outlet of three-cell standard box culvert along Marom creek beneath Bruxner highway B60 at Wollongbar NSW, Australia, on 15 March 2018

Figure 1.1 Photographs of standard box culverts

(C) Culvert along the Enshu coast, Japan, on 21 November 2008

(D) Culvert inlet along Le Gouessant river at Lamballe, France, on 26 June 2019

(E) Multicell masonry culvert inlet along the Flora river at Pléneuf-Val-André, France, on 27 June 2019 – construction second half of 19th century

Figure 1.1 (Continued)

(A) Submerged culvert road embankment in the Coomera river catchment QLD on 31 March 2017 following tropical cyclone Debbie – runoff from right to left

(B) Culvert inlet operation during Norman creek flood on 21 December 2001, Greenslopes QLD, Australia – culvert beneath Cornwall Street

Figure 1.2 Culvert operation in eastern Australia

(C) Culvert operation (barrel entrance) during Norman creek flood on 20 May 2009, Greenslopes QLD, Australia – culvert beneath Ridge Street

Figure 1.2 (Continued)

flow properties (i.e. the bulk velocity V_{mean} and water depth d). The approach encompasses a fair level of empiricism: "this simple 1D approach is clearly problematic" (Morvan *et al.*, 2008, p. 192). In relation to upstream fish passage, by far a most pertinent flow property is the velocity distribution in the vicinity of solid boundaries, given that small fish predominantly swim upstream next to the bottom corners and sidewalls (Blank, 2008; Jensen, 2014; Katopodis and Gervais, 2016; Wang *et al.*, 2016a; Cabonce *et al.*, 2019). A complete characterization of the velocity field requires a detailed investigation, which may be undertaken physically in the laboratory, numerically using computational fluid dynamics (CFD), and possibly theoretically for a few simplistic cases. Laboratory measurements must be based upon a large number of data points to characterize the main stream, boundary regions (i.e. next to bed and walls), and secondary flows, for example, more than 250 to 300 sampling points per cross-section for a given flow rate, as in the studies of Wang *et al.* (2018) and Cabonce *et al.* (2018, 2019). Although the implementation of complex three-dimensional (3D) CFD models in aeronautics and industrial flows has become common (Roache, 1998; Rizzi and Vos, 1998), the application of this approach is much more recent in open channel flows, with inherent difficulties in applying 3D CFD to free-surface flows, for example, the air–water interface, complicated geometry, and roughness definition (Rodi *et al.*, 2013; Khodier and Tullis, 2018; Zhang and Chanson, 2018). Appendix D discusses more specifically the application of 3D CFD to fish-friendly box culvert barrel modeling.

1.2   Impact of road crossings

Freshwater fish species constitute about one-quarter of all living vertebrates and are considered an at-risk group due to deleterious habitat impacts. In Australia, for example, there are about 250 freshwater fish species, with approximately 30% listed as threatened under state and commonwealth legislations (Allen *et al.*, 2002; Lintermans, 2013). The negative effects of river crossings on freshwater fish species have been well documented in the literature (Warren and Pardew, 1998; Briggs and Galarowicz, 2013). Culvert structures create physical or hydrodynamic barriers that often prevent or reduce access to essential breeding and feeding habitats. The direct consequences of losing access to and fragmentation of river habitats on fish encompass reduced recruitment, restricted range size, and changes in fish population composition (Dynesius and Nilsson, 1994; O'Hanley, 2011). Apart from impeding fish passage, road crossing barriers can act in other disruptive ways. Examples include modified suspended and bed load sediment transport; changes in river substrate composition, morphology, and nutrients; and modification of the large woody debris supply (Hotchkiss and Frei, 2007). The resulting environmental changes can extend along the river reaches in both the downstream and upstream directions, including the creation of conditions potentially favorable to the establishment and development of non-native invasive species (Olson and Roy, 2002; Milt *et al.*, 2018). The end result could be a reorganization of the riverine biophysical structure, most often associated with a reduction in the numbers and diversity of native fish species (O'Hanley, 2011).

In natural rivers and hydraulic structures, including culverts, water in motion is turbulent. Turbulent flows are characterized by an unpredictable pseudo-random behavior, strong mixing properties, and a broad spectrum of length and time scales (Adrian and Marusic, 2012). The fluid particles move in very irregular paths, causing an exchange of momentum from one portion of the fluid to another. Although the turbulence may be analyzed in terms of the statistical properties of the velocity components, the turbulence scales are of interest in addition to the turbulence intensity and statistical moments of the turbulent velocity fluctuations. Turbulent vortical structures span over a wide hierarchy of scales, which are all important to turbulent flow science. Despite the recent advances in the hydrodynamics of culverts, there remains a gap between our knowledge of the characterization of turbulence and our understanding of its role on biotic communities. Leading scholars stressed the challenge, specifically the need for "a better understanding of the relationship between the turbulent properties and [their] influence on individual organisms and ecological communities . . . to effectively integrate hydraulically realistic information with ecological data" (Maddock *et al.*, 2013, p. 432). Several researchers have further pointed to the absence of standardized fish swimming tests and data interpretation relevant to engineering design (Kemp, 2012; Katopodis and Gervais, 2016) and the "inconsistent metrics in the published literature" (Kemp, 2012, p. 404). Yet there is a physically based relationship between the local longitudinal velocity and the (mechanical) power and energy required by fish to swim upstream against the discharge (Wang and Chanson, 2018a). This aspect is discussed at more length in Chapter 3 and Appendix C. Let there be no *qui pro quo*: a sound linkage between biology and engineering is a fundamental requirement to advance our understanding of fish-friendly hydraulic structure design.

Given the deleterious environmental problems created by road crossings, it is not surprising that various culvert design guidelines have been developed to facilitate upstream fish passage in culverts (see Table 1.1), albeit not always successfully. Recent field and laboratory works have yielded markedly different recommendations (Table 1.2). Relatively little work has been published regarding the development of systematic design methods for cost-efficient, fish-friendly culverts to deliver continuous fish connectivity over wide geographic areas

Table 1.1 Design guidelines for freshwater fish passage in standard culverts

Reference	Country and region	Targeted fish species	Design criteria	Flow conditions
Fairfull and Witheridge (2003)	Australia		Depth > 0.2–0.3 m V_{mean} < 0.3 m/s for d < 0.5 m	Smooth culvert
Bates et al. (2003)	USA, Washington	Trout, pink salmon, chum salmon, chinook, coho, sockeye, steelhead	Depth > 0.30 m V_{mean} < 0.61 to 1.83 m/s	Smooth culvert
Cahoon et al. (2007)	USA, Montana	Yellowstone cutthroat trout (Oncorhynchus clarkii bouvieri), rainbow trout (Oncorhynchus mykiss)	V_{mean} < 1.9–2.7 m/s	Box culvert geometry
Kilgore et al. (2010); Schall et al. (2012)	USA		Minimum water depth for Q > Q_{min} Maximum bulk velocity for Q < Q_{high}	Q_{min} < Q_{high} < Q_{des}
Courret (2014)	France	Trout, European bullhead (Cottus gobio), brook lamprey (Lampetra planeri), spined loach (Cobitis taenia), common minnow (Phoxinus phoxinus), eel, crayfish	Baffles, macro-roughness	From drought to two to three times the mean annual discharge
DWA (2014)	Germany	European species incl. barbel, brown trout, eel, grayling, salmon, etc.	V_{mean} < U_{fish} Depth > 2.5 × Fish height Baffles/crossbars, macro-roughness	Q_{330} < Q < Q_{30} each year
Chanson and Leng (2019)	Australia	Small-bodied fish species and juvenile of larger fish (L_f < 100 mm, U_{fish} < 0.6 m/s)	LVZ: 0 < V_x < U_{fish} A ≥ 15% for Q ≤ Q_T = 0.1 × Q_{des} DL ≥ 35 mm	Smooth box culvert Q ≤ Q_T < Q_{des} Q_T = 0.1 × Q_{des}

Notes: A: relative low-velocity-zone (LVZ) area; DL: bottom corner LVZ area; L_f: total fish length Q: water discharge; Q_T: threshold discharge; Q_{30}: flow rate occurring no more than 30 days per year; Q_{330}: flow rate occurring no more than 330 days per year; q: unit discharge; S_o: bed slope; U_{fish}: characteristic fish swimming speed; V_{mean}: bulk velocity; V_x: local fluid velocity; z: vertical elevation above the invert.

Table 1.2 Observations and derived recommendations for freshwater fish passage in standard culverts

Reference	Country and region	Targeted fish species	Design criteria	Flow conditions	Type of study
Chorda et al. (1995)	France		Baffles	$S_o = 0.01$ to 0.05	Laboratory work
Gardner (2006)	USA, North Carolina	Bluehead chub (Nocomis leptocephalus), redbreast sunfish (Lepomis auritus), Johnny darter (Etheostoma nigrum), bluegill (Lepomis macrochirus), margined madtom (Noturus insignis), swallowtail shiner (Notropisprocne)	$V_{mean} < 0.55$ m/s	Smooth culvert	Laboratory work Box culvert geometry
Blank (2008)	USA, Montana	Yellowstone cutthroat trout (Oncorhynchus clarkii bouvieri)	$V_x(z = 0.06$ m$) < 1$ to 2 m/s $q < 0.4$ to 0.57 m²/s	Base flow: 0.28 m³/s $S_o = 0.02$ to 0.05	Field observations Box culvert geometry
Monk and Hotchkiss (2012)	USA, Utah	Leatherside chub (Lepidomeda aliciae), speckled dace (Rhinichthys osculus)	$V_x(z = 0.02$ m$) < U_{fish}$	$L_{barrel} \sim 20$ m $0.5 < Q < 1.6$ m³/s $0.073 < q < 0.24$ m²/s	Field observations Box culvert geometry

Notes: d: water depth; L_{barrel}: barrel length; Q: water discharge; S_o: bed slope; V_{mean}: bulk velocity; V_x: local longitudinal velocity; z: vertical elevation above the invert.

(Leng *et al.*, 2019). In most cases, the design methods have focused predominately on baffle installation and boundary roughening along the culvert barrel invert to slow down the water velocity, although the additional flow resistance can reduce drastically the culvert discharge capacity for a given afflux (Larinier, 2002; Olsen and Tullis, 2013). Such a reduction in culvert capacity markedly increases the total cost of the culvert for a design discharge and maximum acceptable afflux. Only a few studies examined robust engineering-based methods (e.g. Papanicolaou and Talebbeydokhti, 2002; Hotchkiss and Frei, 2007; Olsen and Tullis, 2013).

Existing culvert guidelines for upstream fish passage are typically based upon a number of simple criteria, including bulk velocity and minimum water depth (Table 1.1). In Table 1.1, one could note the restrictive nature of many guidelines. Most are unsuitable for the passage of weak-swimming fish and yield expensive culvert structures. Figure 1.3 illustrates a multicell box culvert recently renovated to provide fish passage, albeit using a complicated and relatively expensive approach.

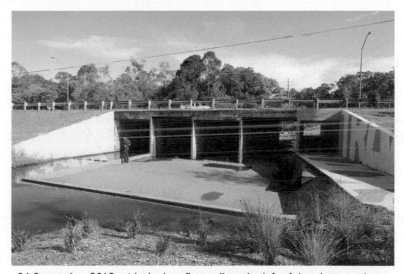

(A) Inlet on 26 September 2018, with the low-flow cell on the left of the photograph

(B) Outlet on 21 April 2018 with the fish-friendly rock ramp on the right of the photograph

Figure 1.3 Retrofitted box culvert outlet with fish-friendly low-flow cell along Slacks creek in Logan QLD, Australia, in 2018, looking upstream at the outflow channel (foreground right) and culvert outlet – the culvert inlet, left barrel cell, and outflow channel were retrofitted to facilitate fish passage at low flow, at an estimated cost of A$125,000

1.3   Culvert design, construction, and operation

A thorough hydraulic design requires considerations of some aspects of hydrology, construction, inspection, and maintenance. During the design stages, discharge estimates are required for the evaluation and determination of the culvert capacity. Construction methods and schedules should consider the hydrology of the catchment, the hydraulic characteristics of the stream, and the impact of culvert construction on its environment. During culvert inspections, problems and deficiencies must be recognized, documented, and drawn to the attention of the relevant specialists. The maintenance and services of culverts must include the removal of sediments and debris obstructing the waterway. The best designs of culvert structures are often obtained when there are close collaboration and interactions between the design and construction teams, as well as the asset operator. In many instances, the initial designs of foundations, culvert barrel, and embankments may be modified to some degree to address construction constraints, schedule, and overall costs. The following paragraphs are intended for standard box culverts along small to medium streams beneath two-lane roads. Longer or wider culvert structures may require special requirements for design, construction, inspection, or maintenance that are outside of the scope of this monograph.

1.3.1   Catchment hydrology

Engineering hydrology deals with the relationship between rainfall and runoff discharge. Figure 1.4 illustrates the observed response of a small catchment to a relatively intense storm event. At the beginning of the rainstorm event, the rainfall runoff was initially intercepted by the vegetation of the dry catchment and infiltrated into the ground, and a period of time passed before the creek began to rise. Once the catchment, entirely or partially, became saturated, the rainfall started to contribute directly to the river discharge. In terms of

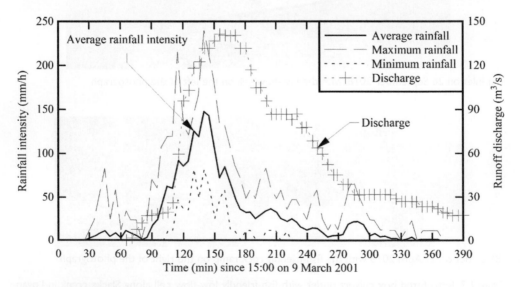

Figure 1.4 Rainfall intensity across the catchment and runoff discharge hydrograph during a flood event on 9 March 2001 at Holland Park East station in upper Norman creek catchment, QLD, Australia (catchment area: 8.5 km²) – data: Yu *et al.*, (2007)

hydraulic engineering, the design of a culvert is based upon predicted catchment outflows for specified design storms. The discharge estimates are used to calculate the dimensions of the required culvert structure and to assess its impact on the upstream and downstream catchments (e.g. in terms of the afflux at design flow rate).

Stream flow analyses may be divided into three categories: (a) statistical frequency analyses of gauged stream flow data, (b) runoff modeling based upon rainfall data as input, and (c) empirical methods (e.g. the rational method). Whatever the method(s) used to derive the design discharge, the engineers must consider the risks of losses of life and property associated with potential exceedance of the design discharge selection and act accordingly in a conservative and professional manner. In Australia, for example, the Australian rainfall and runoff (ARR) provides a national guideline document, data, and software suite for the estimation of design flood characteristics (Ball *et al.*, 2016).[3]

1.3.2 Construction and soil mechanics considerations

The safety of the embankment depends to a large extent upon the stability of the culvert structure. Any leakage or failure of the culvert conduit may cause openings through the embankment, which may progressively develop until its partial or complete failure. Percolation is also possible along the contact surface between the culvert's outer shell and the earthfill embankment, resulting in serious damage. Another challenge is the possibility of structural collapse of the conduit, which would result in the failure of the embankment. There are different types of culvert construction, and Figure 1.5 presents the main types of box culvert units. Figure 1.6 illustrates the details of prototype structures.

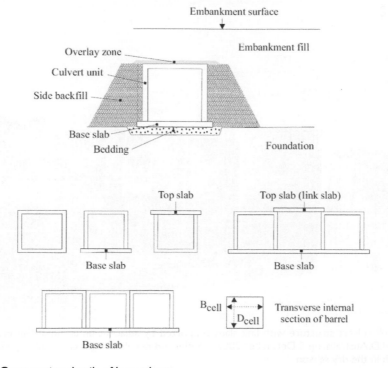

Figure 1.5 Construction details of box culverts

(A) Multicell inverted U-shaped barrel, in Salisbury QLD, Australia, on 19 July 2018 – culvert inlet along Stable Swamp creek beneath Musgrave Road

(B) Multicell culvert structure with link slabs supported by adjacent units, along Gin House creek, Carrara QLD, Australia, on 5 December 2007 – outlet and details of right cell used as pedestrian and cyclist path in the dry season

Figure 1.6 Photographs of different types of culvert construction

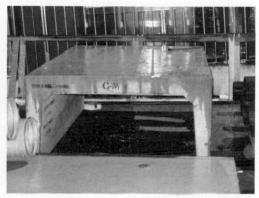

(C, Left) One-piece unit, along Norman creek QLD, Australia

(D, Right) Precast concrete box element (U shaped crown unit) – the 1200 mm × 600 mm element weighed 2190 kg

(E) Stone masonry box culvert inlet at La Cave in Matignon, France, on 29 June 2019 – construction from second half of 19th century with piers made of granite stones

Figure 1.6 (Continued)

In practice, the conduit joints must be watertight to prevent leakage into the surrounding earthfill. Joints between precast concrete units should be butt joints (Australian Standard, 2010, 2013). With prefabricated units, the methods of bedding and backfilling the conduit should preclude unequal settlement and ensure an uniform load distribution on the foundation. Extreme care should be taken to secure a tight contact between the fill and the conduit's outer surface. This is critical to prevent percolation along the culvert unit's outer shell, as well as to ensure that the fill develops a lateral restraint, preventing excessive stresses on the shell (USBR, 1987). For a long culvert barrel, cutoff collars may be considered to reduce seepage and the risks of piping around the conduit shell, although there are other mitigation methods. Cutoff collars are typically located between joints in the conduit.

With precast concrete units, the culvert must be set carefully on a good foundation, often bedded in concrete. The concrete base prevents seepage along the underside of the conduit and supports the boxes both laterally and longitudinally. With U-shaped units and one-piece culvert units, their base must be supported by a bed zone layer (Australian Standard, 2010, 2013) (Fig. 1.5).

Finally the embankment may be overtopped during exceptional flood events, and it must be designed accordingly (CIRIA, 1987).

1.3.3   Inspection

After completion, culvert structures must be inspected systematically at regular intervals. Timely maintenance works may reduce the risk of severe damage and failure of a culvert, including during an extreme event. For example, on 8 June 2007, five people were tragically killed when their car was swept away as a culvert beneath the old Pacific Highway near Gosford NSW (Australia) collapsed during a major flood of Piles creek (Clarke *et al.*, 2019); lack of maintenance was blamed by the coroner.

Evidence of hydraulic issues may occur in various sub- and super-structure components, for example, road cracks might indicate differential settlement of a barrel cell. When foundation problems are suspected, geotechnical, structural, and hydrotechnical experts must be involved as part of the inspection and assessment process. If the inspection suggests the possibility of dangerous conditions, asset owners and managers must be informed, and embankment (e.g. road) closure might be necessary so that remedial work can be started promptly.

1.3.4   Maintenance

Culvert maintenance regroups work done on a regular basis as well as in an emergency situation to maintain the integrity of the culvert and embankment and protect against future flood events. Maintenance must be closely linked to inspection and design, as feedback from inspection and maintenance teams can reduce problems in future designs. The maintenance of culvert structures encompasses the removal of sediment and debris obstructing the waterway, in particular, upstream of the culvert and in the culvert inlet and barrel (Fig. 1.7). Recurring issues may encompass scour and erosion during flood events. These problems

(A, Left) Inlet of three-cell box culvert blocked by large tree logs and branches along Marom creek beneath Bruxner Highway B60 (Wollongbar NSW, Australia) on 18 October 2016

(B, Right) Box culvert inlet blocked by a large amount of debris in Aachen, Germany, in May 2018

(C) Sedimentation in a box culvert in Salisbury QLD, Australia – sediment removal from culvert along Rocky Water Hole beneath Gladstone Street on 19 July 2018

Figure 1.7 Obstruction of culverts

should be considered during the design stages and could sometimes be eliminated during construction (e.g. with the addition of an apron in the outlet).

A related challenge is the frequency of embankment overtopping in relation to the traffic on the embankment (e.g. when the traffic increases with a denser population in the area). The hydraulic implications should then be investigated carefully (e.g. raising the embankment could cause adverse impacts with liability implications).

1.4 Structure of the monograph

In this monograph, new guidelines are proposed for fish-friendly, multicell box culvert designs. The focus is on box culverts, with a novel approach based upon three basic concepts: (a) the culvert design is optimized for fish passage for small to medium water discharges and for flood capacity at larger discharges, (b) low-velocity zones are provided along the wetted perimeter in the culvert barrel and quantified in terms of a fraction of the wetted flow area where the local longitudinal velocity is less than a characteristic fish speed linked to swimming performances of targeted fish species, and (c) a smooth box culvert barrel is designed without appurtenance. The approach relies upon an accurate, physically based knowledge of the entire velocity field in the barrel, specifically the longitudinal velocity map, since a number of fish behavior observations showed that weak-swimming fish swim in low-velocity zone (LVZ) boundaries (Appendices C and F). The targeted fish species are small-bodied (less than 100 mm long) and juvenile native fish species, although the approach and methodology are general and applicable to other fish guilds.

The monograph develops basic hydraulic engineering guidelines for the design of fish-friendly standard box culverts. The concepts and their application are relevant to most box culvert structures and fish species. Chapter 2 presents current practices for the hydraulic engineering design of standard box culverts, aimed to optimize the culvert structure for its design flow conditions. Chapter 3 discusses fish passage in a standard box culvert barrel. Chapter 4 details a new approach for the hydraulic engineering design of fish-friendly standard box culverts. Chapter 5 shows a complete design application. Chapter 6 discusses practical considerations relevant to fish-friendly box culvert designs.

Appendix A develops a detailed glossary of technical terms. Appendix B presents hydraulic engineering calculations of flood plains, as well as hydraulic engineering calculations of standard culvert operation with outlet control for less-than-design discharges. Appendix C discusses the intricacies of physical modeling of fish passage in culverts, including details of recent laboratory studies of fish swimming in a near-full-scale box culvert barrel channel. Appendix D presents the CFD modeling of fish-friendly standard box culverts. Appendix E develops a few alternatives to improve the upstream passage of small-body-mass fish, including retrofitting. Further illustrations are presented in Appendix F in the form of a series of movies of box culvert operation and high-speed video movies of fish swimming. The monograph ends with an author index, a subject index, and a list of bibliographic references.

Notes

1 That is, an excessive vertical drop at the culvert exit.
2 The design discharge is derived from a system analysis of the catchment hydrology and hydraulics in relation to the purpose of the culvert (Chapter 3). At design flow, the embankment should not be overtopped and the barrel should operate as a free-surface flow.
3 See the Internet bibliography at the end of the list of references.

Chapter 2

Hydraulic engineering design of standard box culverts – current practice

2.1 Presentation

During the design of a culvert, the primary constraints are (a) passing the design flood while maintaining an appropriate level of free-board below the embankment roadway and (b) keeping the total cost to a minimum and the afflux as small as possible (Herr and Bossy, 1965; Chanson, 1999a, 2004). The afflux[1] is the rise in upstream water level during a flood, caused by the presence of the culvert structure; it constitutes a quantitative measure of the upstream flooding induced by the culvert. Within current engineering design practices, the basic hydraulic characteristics of a culvert are its design discharge Q_{des} and the maximum acceptable afflux h_{max} at design flow. The design discharge and corresponding water level in the natural stream in the absence of a culvert structure are deduced from the hydrological and hydraulic engineering data of the site in relation to the purpose of the culvert. The afflux must be minimized to reduce flooding in the upstream catchment, as well as the risks of embankment overtopping (Fig. 2.1). The hydraulic design is fundamentally an optimization between the discharge capacity, afflux, and total cost of the culvert structure. Most culverts are designed to operate as open channel systems for discharges up to and including the design discharge Q_{des}. While a key objective is to keep the cost of the culvert to a minimum, some consideration must be given to keep the head loss and afflux small and to avoid downstream scour at the culvert outlet (e.g. using some scour protection measures) (Hee, 1969; QUDM, 2016).

For standard culverts, the traditional engineering design procedure consists of two successive stages. First a complete system analysis must be carried out to ascertain the culvert function(s), the design data, and the design constraints. The first part results in the selection of the design rainfall and runoff event, for example, a 20% annual exceedance probability (AEP) storm, with an estimate of the corresponding design flow rate Q_{des}. The maximum acceptable afflux h_{max} at design discharge is typically selected by the asset owner based upon an assessment of the culvert impact (e.g. on the catchment and embankment). During the second stage, the culvert barrel cross-section area is selected by an iterative procedure in which both inlet control and outlet control calculations are conducted. In hydraulic engineering, inlet control implies that the hydraulic control is located at the entrance of the barrel (e.g. critical flow conditions take place in a barrel with a free-surface inlet) (Fig. 2.2). Outlet control means that the culvert flow is controlled at the outlet (i.e. by the tailwater flow conditions). The optimum size is the smallest barrel size allowing for inlet control operation (Herr and Bossy, 1965; Chanson, 1999a). If the iterative calculations do not converge to a satisfactory solution, the system analysis (i.e. the first stage) must be reconsidered.

(A) Road overtopping on Tuesday, 11 January 2011, afternoon

(B) Dry road on 17 January 2011

Figure 2.1 Haigslee-Fernvale Road culvert embankment overtopping – runoff flow direction from left to right

2.2　Design method and procedure

The barrel size is selected by a test-and-trial procedure. The key output of the iterative calculations is the minimum internal barrel width B_{min} to achieve inlet control. The construction cost may be optimized using a multicell culvert of precast rectangular box elements, and the number of cells N_{cell} becomes the basic output (Fig. 2.2). Other relevant parameters include the bed slope S_o and the tailwater depth d_{tw}. The bed slope is directly proportional to the natural drop in bed elevation along the culvert length, which would be equal to the maximum acceptable head loss for a zero afflux design. Indirectly the bed slope affects the tailwater depth. The tailwater depth is linked to the topography of the downstream catchment (e.g. shape, longitudinal slope, boundary roughness, and possibly tailwater effects) (Appendix B). In a number of cases, the tailwater depth is equal to or close to the uniform equilibrium flow depth in the downstream flood plain for the design discharge, assuming a mild slope. More generally, basic hydraulic calculations are conducted assuming implicitly a mild slope, for which both gradually varied flow and uniform equilibrium flow conditions correspond to a subcritical flow motion (Appendix B).

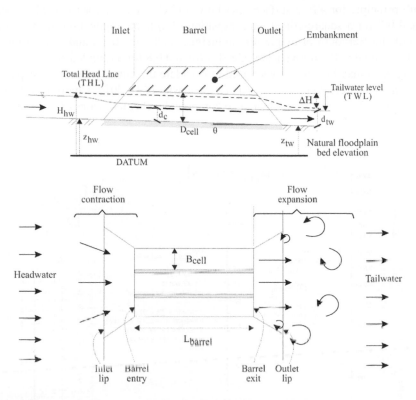

Figure 2.2 Definition sketch of a standard multicell box culvert operation

Finally a multicell culvert structure is basically designed assuming a number N_{cell} of identical cells. Implicitly the calculations assume the same water discharge in each cell (i.e. $Q_{cell} = Q/N_{cell}$) for all discharges.

2.2.1 Discharge calculations

The discharge capacity of a culvert barrel is primarily related to the free-surface flow pattern (Herr and Bossy, 1965; Hee, 1969). When free-surface critical flow takes place in the barrel at design flow, the discharge is fixed only by the entry conditions (i.e. inlet control). With culvert barrels flowing full, the discharge is controlled by the downstream flow conditions (i.e. outlet control).

For relatively short box culverts in which the discharge is controlled by the inlet conditions, the discharge capacity may be estimated based upon theoretical considerations (Henderson, 1966):

$$\frac{Q_{des}}{B_{min}} = C_D \times \frac{2}{3} \times \sqrt{\frac{2}{3} \times g} \times (H_{hw} - z_{hw})^{3/2} \qquad \text{for} \quad \frac{H_{hw} - z_{hw}}{D_{cell}} < 1.2 \qquad (2.1)$$

$$\frac{Q_{des}}{B_{min}} = C \times D_{cell} \times \sqrt{2 \times g \times (H_{hw} - z_{hw} - C \times D_{cell})} \qquad \text{for} \quad \frac{H_{hw} - z_{hw}}{D_{cell}} > 1.2 \qquad (2.2)$$

where B_{min} is the internal barrel width,[2] D_{cell} is the internal barrel height, and the subscript hw refers to the headwater conditions (Fig. 2.2). Equation (2.1) is a direct application of the

Bernoulli principle for a free-surface inlet flow, and C_D equals 1 for rounded vertical inlet edges and 0.9 for a square-edged inlet. Equation (2.2) is an application of the sluice gate solution of the Bernoulli principle for submerged inlet conditions and free-surface barrel flow, with C equal to 0.6 for a square-edged soffit and 0.8 for a rounded soffit.

Nomographs may be alternatively used to calculate the barrel's discharge capacity (e.g. USBR, 1987; Concrete Pipe Association of Australasia, 1991, 2012; Chanson, 1999a, 2004). Figures 2.3 and 2.4 present nomographs to compute the characteristics of box culverts with

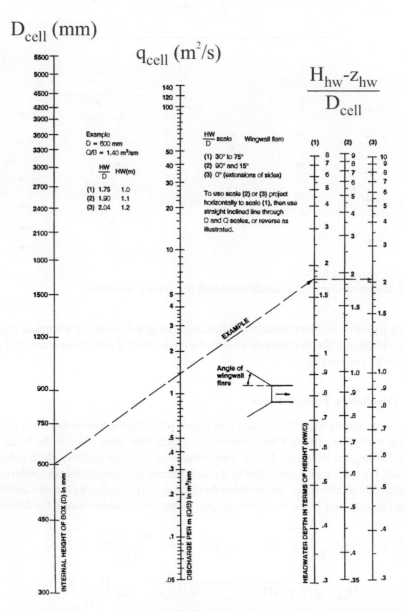

Figure 2.3 Hydraulic calculations of dimensionless headwater total head above invert bed $(H_{hw} - z_{hw})$ / D_{cell} for box culverts with inlet control (Concrete Pipe Association of Australasia, 2012, p. 31)

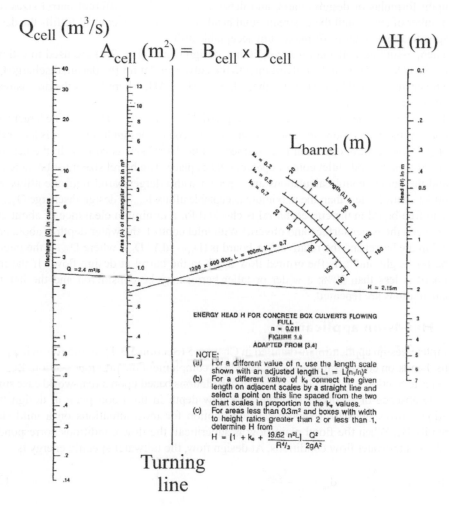

Q_{cell} (m³/s)

A_{cell} (m²) = B_{cell} × D_{cell}

ΔH (m)

L_{barrel} (m)

Turning line

Figure 2.4 Hydraulic calculations of total head losses for concrete box culverts flowing full (Concrete Pipe Association of Australasia, 2012, p. 33)

an inlet control and drowned barrel,[3] respectively. Further, a number of hydraulic design software packages, such as HEC-RAS and SWMM, incorporate the culvert design equations.

2.2.2 Hydraulic design procedure

The calculations of the culvert barrel size are iterative (Herr and Bossy, 1965; Concrete Pipe Association of Australasia, 1991, 2012; Chanson, 1999a). The iteration steps include:

1   First the barrel and element dimensions are selected (e.g. barrel length L_{barrel}, precast box internal dimensions) (B_{cell}, D_{cell}).
2   Next, assuming inlet control conditions, the estimate of the upstream total head $H_{hw}^{(ic)}$ corresponding to the design discharge Q_{des} is undertaken. The calculations may be based

upon formulas or design charts, and these are repeated for different barrel sizes (e.g. number of cells) until the upstream total head $H_{hw}^{(ic)}$ with inlet control fulfills the design specifications in terms of maximum acceptable afflux h_{max}.

3   Then, assuming outlet control conditions, design charts (Fig. 2.4) are used to estimate the head loss ΔH from the culvert inlet to the culvert outlet for the design discharge Q_{des}. The upstream total head $H_{hw}^{(oc)}$ is then $H_{hw}^{(oc)} = H_{tw} + \Delta H$, where H_{tw} is the downstream, tailwater total head.

4   The inlet and outlet control results are compared: $H_{des} = H_{hw}^{(ic)} >/< H_{hw}^{(oc)}$?[4] The larger value controls the culvert flow operation. When the inlet control design head $H_{hw}^{(ic)}$ is larger than $H_{hw}^{(oc)}$, inlet control operation is confirmed and the barrel size is correct. In the negative, $H_{hw}^{(ic)} < H_{hw}^{(oc)}$ and outlet control operation takes place. The barrel size must be increased, and the outlet control calculations are repeated with a larger barrel until the afflux with outlet control is less than the maximum acceptable afflux h_{max} at design discharge Q_{des}.

5   The free-board in the culvert barrel is checked for a minimum clearance of about 20% between the water surface and obvert.[5] With inlet control, the water depth is about critical in the barrel and the relative free-board is $(D_{cell} - d_c) / D_{cell}$, where D_{cell} is the internal barrel height and d_c is the critical flow depth in the barrel at design flow.[6] If the free-board is less than 20%, a wider or taller barrel may be considered and the iterative calculations are repeated.

2.3   Hands-on application

A complete design application is detailed in Chapter 5 (Section 5.3.1). In the following paragraphs, hands-on considerations are developed to complement the previous section. Readers may refer to Section 5.3.1 for an example of calculations based upon a real-world case study.

In the absence of a culvert structure, the flow depth in the flood plain at design flow conditions corresponds to a subcritical flow motion for most situations on a mild slope (Appendix B). When the flood plain flow is subcritical, the flow conditions correspond to the culvert's tailwater flow conditions. At design flow, the tailwater specific energy is:

$$E_{tw} = d_{tw} + \frac{V_{tw}^2}{2 \times g} = d_{tw} + \frac{Q_{des}^2}{2 \times g \times A_{tw}^2} \tag{2.3}$$

where A is the river channel flow cross-section area for the water depth d and the subscript tw refers to the tailwater conditions.

With zero afflux, the upstream and downstream flow depths are both equal to d_{tw}, and the culvert barrel will operate with inlet control conditions. Outlet control calculations are typically not required then.

With a maximum acceptable afflux $h_{max} > 0$, the upstream flow depth is $d_{hw} = d_{tw} + h_{max}$ and the corresponding upstream-specific energy is:

$$E_{hw} = H_{hw} - z_{hw} = d_{hw} + \frac{V_{hw}^2}{2 \times g} = d_{tw} + h_{max} + \frac{Q_{des}^2}{2 \times g \times A_{hw}^2} \tag{2.4}$$

where the subscript hw refers to the headwater conditions.

First the inlet control calculations are conducted. The input variables are the internal barrel height D_{cell}, the upstream-specific energy E_{hw}, and possibly the inlet wingwall configuration.

The output is the discharge per unit width q_{des} (Fig. 2.3). The minimum internal barrel width is then $B_{min} = Q_{des}/q_{des}$. For a multicell culvert structure with identical cells of internal width B_{cell}, the number of cells N_{cell} is the smallest integer value larger than B_{min}/B_{cell}. Conversely, for a structure made of N_{cell} identical barrel cells, the design chart (Fig. 2.3) gives the expected afflux.[7]

Second, the outlet control calculations are performed. The input data are the barrel's internal cross-section area A_{barrel}, the barrel length L_{barrel}, an entrance loss coefficient (e.g. $k_e = 0.5$), and the water discharge Q_{des}.[8] The calculation output is the head loss ΔH (Fig. 2.4). The upstream total head with outlet control is then $H_{hw}^{(oc)} = H_{tw} + \Delta H$. In the first approximation, the afflux may be estimated as $\Delta H - L_{culv} \times S_o$, where L_{culv} is the total culvert length measured from inlet lip to outlet lip, and S_o is the longitudinal bed slope ($S_o = \sin\theta$).

If the afflux is greater with outlet control than for inlet control operation, the barrel size must be increased. The outlet control calculations are repeated with a larger barrel size until the outlet control afflux is smaller than the maximum acceptable afflux h_{max} at design flow conditions (Herr and Bossy, 1965, pp. 5–17; Chanson, 2004, p. 454). Inlet control calculations do not need to be checked, since the smaller size was satisfactory for this control, as determined under the first step.

Finally the free-board in the barrel is checked.

2.4 Practical considerations

The current hydraulic design of a standard box culvert is an optimization process for the design flow conditions. Consideration for nondesign flow conditions is limited.

In practice, the design engineers have a responsibility to ensure that a culvert operates safely for a broad range of flow conditions (Cottman et al., 1980; Schall et al., 2012; QUDM, 2016). Damage (e.g. scouring, piping, breaching) to the embankment and to the downstream river bed may occur in several cases:

- The apron is too short and/or too shallow to prevent bed scour
- Flow conditions are larger than design flow conditions, leading to embankment overtopping
- Inlet blockage by debris
- Sediment siltation and buildup in culvert barrel
- Unusual flood event during construction periods
- Poor construction of the barrel, inlet or outlet
- Poor shapes of the inlet and outlet, or misalignment of the barrel in relation to the stream flow direction, resulting in poor discharge capacity[9]
- Wrong dimensions of the barrel

Current engineering design practices have been developed for the reference flow conditions (i.e. design flow conditions). For discharges larger than the design discharge, it may be acceptable to tolerate some erosion and damage. However, it is essential that the stability and integrity of the embankment are ensured. For flow rates smaller than the design discharge, perfect performances are expected: that is, (a) the culvert must operate safely and (b) there must be no maintenance issues. These objectives are achieved by a correct design of culvert barrel dimensions, a correct design of inlet and outlet sections to guide the flow into and out of the culvert barrel, and provision of a downstream apron to prevent downstream bed scour, if required.

Notes

1 A glossary of technical terms is developed in Appendix A.
2 For a multicell box culvert, $B_{min} = N_{cell} \times B_{cell}$.
3 At design discharge ($Q = Q_{des}$), a culvert operating with outlet control often presents a drowned barrel.
4 Inlet and outlet control results may also be compared in terms of the corresponding afflux. However, the culvert hydraulics are driven by energy considerations, and a comparison in terms of the head-water total head H_{hw} is more pertinent.
5 The obvert or soffit is the roof of the culvert barrel (Appendix A).
6 Critical flow conditions correspond to a maximum discharge per unit width in the barrel, allowing for a minimum internal barrel width.
7 The expected afflux is typically less than h_{max} for a multicell structure, since $N_{cell} \times B_{cell} > B_{min}$.
8 For a multicell culvert structure, outlet control calculations may be conducted for a single cell. The input is then the cell's internal cross-section area $A_{cell} = D_{cell} \times B_{cell}$, the barrel length L_{barrel}, an entrance loss coefficient (e.g. $k_e = 0.5$), and the water discharge $Q_{des} = Q_{des}/N_{cell}$ per cell.
9 A competent professional engineer would always attempt to align the culvert barrel to the waterway course, sometimes leading to a skewed barrel. Although the skewness adds to the cost, it improves drastically the hydraulic efficiency of the structure, in turn reducing the total cost. When the culvert barrel is built at an angle to the water flow, flow separation may take place in the inlet, barrel, and outlet, and the previous equations and nomographs might not be applicable.

Chapter 3

Fish passage in standard box culverts

3.1 Presentation

In terms of upstream fish passage in standard culverts, the most relevant parameters include the culvert barrel dimensions, cross-sectional shape and invert slope, the water discharge, the fluid dynamics properties in the barrel, and the targeted fish species. Box culverts are generally considered more effective for fish passage than circular pipes (Briggs and Galarowicz, 2013). The behavioral response by fish species to culvert dimensions and hydrodynamic flow conditions, including turbulence, may play a role in their ability to successfully pass the culvert. The length of the barrel may be another important factor, with increasing fish passage limitations, with increasing barrel length increases for some fish species. While there are a broad range of culvert designs resulting in a wide diversity in turbulent flow patterns, there are still ongoing discussions regarding how the flow turbulence characteristics might interplay with fish passage and fish behavior (Liu *et al.*, 2006; Wang and Chanson, 2018b). A few seminal works debated the most relevant turbulence characteristics to assist fish passage (Pavlov *et al.*, 2000; Hotchkiss, 2002; Crowder and Diplas, 2002; Nikora *et al.*, 2003). Simply, the fish–turbulence interactions are extremely complicated, and naive "turbulence metrics cannot explain all the swimming path lines or behavior" (Goettel *et al.*, 2015, p. 239).

The interplay between fish and turbulence encompasses a broad range of relevant length and time scales and is scale dependent (Lupandin, 2005; Webb and Cotel, 2011; Wang and Chanson, 2018a). Furthermore, turbulence modulation by fish swimming cannot be ignored. A broad range of turbulent flow properties constitute the determining factors characterizing the ability of the targeted fish species to pass the culvert, especially for small-bodied, weak-swimming fish species, including juveniles of larger-bodied fish. A seminal discussion emphasized the role of secondary flow motion and "the importance of performing three-dimensional turbulent flow measurements to precisely identify the effects of secondary flows on fish motion" (Papanicolaou and Talebbeydokhti, 2002, p. 548). Next to a sidewall, the channel flow is retarded, and complicated flow patterns may develop (e.g. next to the corners of the side and base slabs) (Bradshaw, 1987; Sanchez *et al.*, 2018). The result is some secondary flow motion generated at a right angle to the longitudinal current, inducing a sizable low-velocity zone (LVZ) in a box culvert barrel, as sketched in Figure 3.1.

Recently, field observations and large-size laboratory studies documented fish swimming and behavior in box culvert barrels (Blank, 2008; Jensen, 2014; Wang *et al.*, 2016a; Cabonce *et al.*, 2017, 2018, 2019). All the data indicated that the fish swim preferentially close to the

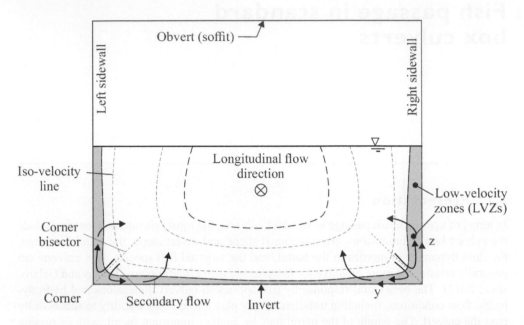

Figure 3.1 Sketch of velocity contour map and secondary current in a box culvert barrel cell looking downstream

sidewalls, particularly in the bottom corners, in regions of low velocity and high turbulence intensity (Fig. 3.1) (also Appendix C). This finding is seminal. In the presence of various types of boundary treatments, the observations showed the "sweet spots" (i.e. regions of slower velocity and high turbulence) that the fish exploit. Figure 3.2 presents photographs of small-body-mass fish swimming next to the bottom corners in box culvert channels, with a range of wall treatments, including smooth boundaries (Figs. 3.2A and 3.2B). Each figure caption includes the water discharge Q, the barrel channel width B, the fish species, and the total length L_f of the fish. Appendix F presents several movies of fish swimming in the bottom corners with different boundary treatments: for example, the movies CIMG1655.mov, CIMG2672.mov, CIMG1647.mov, CIMG2725.mov, and CIMG1651.mov in a smooth box culvert channel; the movies CIMG1497.mov and CIMG2523.mov in a barrel channel with a very rough invert; and the movies CIMG1243.mov, CIMG1268.mov, CIMG1409.mov, and CIMG1410.mov in a box culvert barrel with large asymmetrical roughness.

A key flow region is the bottom barrel corner region, where secondary currents are strong (Fig. 3.1). Figure 3.3 shows the proportion of time spent by small-body-mass fish in a smooth culvert barrel channel for two species. Further data are reviewed in Appendix C. Irrespective of the boundary treatment, detailed observations indicated that studied fish spent two-thirds of their time in the bottom corners and nearly 90% of the time next to the sidewalls and bottom corners altogether (Fig. 3.3). Detailed observations and high-speed movies (Appendix F) suggested that the fish then "waltz dance" with the turbulent coherent structures to minimize acceleration/deceleration and the associated energy consumption (Wang and Chanson, 2018b, p. 27). Simply put, the upstream passage may be successful when the fish use the turbulence and not fight it.

Figure 3.2 Fish behavior in box culvert barrel channels, with flow direction from left to right – (A) Duboulay's rainbowfish (*Melanotaenia duboulayi*) swimming upstream along the right side-wall in a smooth channel (Q = 0.026 m³/s, B = 0.5 m, $L_f \approx$ 60 mm); (B) Juvenile silver perch (*Bidyanus bidyanus*) swimming next to right smooth sidewall in a rough invert channel (Q = 0.026 m³/s, B = 0.5 m, $L_f \approx$ 120 mm); (C) Juvenile silver perch swimming next to left rough sidewall in an asymmetrically roughened channel (Q = 0.026 m³/s, B = 0.5 m, $L_f \approx$ 90 mm); (D) Juvenile silver perch resting upstream of a small corner baffle (Q = 0.056 m³/s, B = 0.5 m, h_b = 0.066 m, $L_f \approx$ 60 mm); (E) Juvenile silver perch behind a small corner baffle (Q = 0.056 m³/s, B = 0.5 m, h_b = 0.133 m, $L_f \approx$ 70 mm); (F) Juvenile silver perch negotiating successfully a small corner baffle (Q = 0.056 m³/s, B = 0.5 m, h_b = 0.066 m, $L_f \approx$ 70 mm)

(F)

Figure 3.2 (Continued)

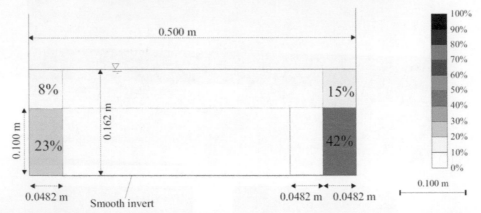

(A) Juvenile silver perch (*Bidyanus bidyanus*) in a smooth rectangular channel with a subcritical flow: $Q = 0.0556$ m³/s, $x = 4$–6.5 m, $d = 0.161$ m, $V_{mean} = 0.69$ m/s, $\theta = 0$

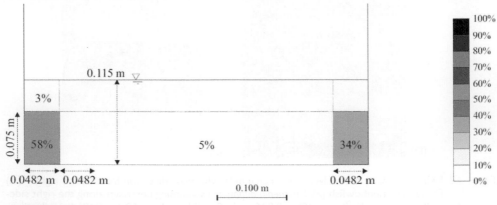

(B) Duboulay's rainbowfish (*Melanotaenia duboulayi*) in a smooth rectangular channel for a subcritical flow: $Q = 0.0261$ m³/s, $x = 4$–6.5 m, $d = 0.11$ m, $V_{mean} = 0.54$ m/s, $\theta = 0$

Figure 3.3 Percentage of time spent by small-body-mass fish within a smooth box culvert barrel channel ($B_{cell} = 0.5$ m) weighted with respect to time – fish swimming tests detailed in Table 3.1

By analogy to competitive swimming and sport physics (Counsilman, 1968; Wang and Wang, 2006; Clanet, 2013), the upstream passage of fish in culverts could further be analyzed in terms of an optimization process. The basic concepts of fish swimming dynamics are the notions that (a) the travel time of the fish in the culvert barrel equals the ratio L_{barrel}/U_{fish} of barrel length to absolute fish speed and (b) the rate of mechanical work exerted by a fish is equal to the thrust times the relative fish speed, and is hence proportional to the cube of the local fluid velocity (Wang and Chanson, 2018a, 2018b). Fish may further adapt their swimming stroke to maximize their efficiency, as observed with competitive swimmers during international and Olympic meets (Kolmogorov and Duplishcheva, 1992; Wei *et al.*, 2014). Such a concept, however, raises questions on the limitations of many fish swim tests in tubes and tunnels (Katopodis and Gervais, 2016) (Appendix C).

Commentary

by Dr. Hang Wang, Sichuan University, State Key Laboratory of Hydraulics and Mountain River Engineering, Chengdu, China

"This is a golden opportunity for your world to communicate with our world. Fisheries scientists would benefit from better understanding of flows and force."
– Jeff Schaeffer, former editor of *Fisheries* (16 May 2017)

Although written for hydraulic engineers, this book highlights the interdisciplinary nature of the challenges raised with the modern design and redevelopment of fish-friendly culverts and the need for involvement of biological expertise in engineering design. Many early studies aiming at better adaptation of fish to manmade structures were carried out with limited cross-disciplinary understanding between hydraulic engineering and biological science, sometimes leading to insubstantial scientific bases associated with, for example, incomplete biological protocols in hydraulic modeling dealing with fish observations or oversimplification of dynamic flow environments in fish behavior studies. The authors' work provides a delicate approach bridging the two fields through quantitative characterization of the coupling between fish and flow kinematics. Such physical insights have not been possible without either advanced experimental and data analysis techniques (fish tracking, high-resolution turbulence measurements, etc.) or engineering researchers understanding biological language and vice versa. Energy consideration is an obvious step forward toward the nature of the fish's response to natural or artificial turbulence. The role of turbulence that is sensed by fish, in comparison to the mean flow properties perceived by fish, is also to be clarified in convincing ways in the future.

3.2   Physical modeling of fish passage: basic dimensional considerations and similitude

A laboratory model is designed to provide reliable predictions of the full-scale hydraulic structure (Henderson, 1966). The physical modeling, including fish swimming, must be based upon the fundamental concepts and principles of similitude. In the dimensional analysis of fluid dynamics, the relevant variables include the fluid properties, physical constants, boundary conditions (including channel geometry), and initial flow conditions (Liggett, 1994; Foss et al., 2007).

Let us consider a steady turbulent flow in a smooth box culvert barrel operating as a free-surface flow (Fig. 3.1). The dimensional analysis gives a series of relationships between the fluid flow properties at a location (x, y, z) and the upstream flow conditions, boundary conditions, and fluid properties:

$$d, \vec{V}, v', p, L_t, T_t, \ldots = F_1\left(x, y, z, B_{cell}, k_s, \theta, d_1, V_1, v_1', \rho_w, \mu_w, \sigma, g, \ldots.\right) \tag{3.1}$$

where d is the flow depth; V is the velocity; v' is a velocity fluctuation; p is the pressure; L_t and T_t are integral turbulent length; time scales, x, y, and z are, respectively, the longitudinal, transverse, and vertical coordinates; B_{cell} is the internal barrel width; k_s is the equivalent sand roughness height of the culvert barrel boundary; θ is the angle between the culvert invert and horizontal; d_1, V_1, and v_1' are, respectively, the inflow depth, velocity, and velocity fluctuation; ρ_w and μ_w are the water density and dynamic viscosity; σ is the surface tension; and g is the gravity acceleration.

The $\prod$-Buckingham theorem states that any dimensional equation with N variables with units encompassing mass, length, and time may be rewritten into an equation with (N-3) dimensionless parameters (Vaschy, 1892; Buckingham, 1914; Rouse, 1938).[1] In turn, Equation (3.1) may be expressed:

$$\frac{d}{d_c}, \frac{V_x}{V_c}, \frac{v_x'}{V_c}, \frac{p}{\rho_w \times g \times d_c}, \frac{L_t}{d_c}, T_t \times \sqrt{\frac{g}{d_c}} \ldots = F_2 \left(\begin{array}{c} \dfrac{x}{d_c}, \dfrac{y}{d_c}, \dfrac{z}{d_c}, \\[2mm] \dfrac{B_{cell}}{d_c}, \theta, \dfrac{k_s}{d_c}, \\[2mm] \dfrac{d_1}{d_c}, \dfrac{V_1}{\sqrt{g \times d_1}}, \dfrac{v_1'}{V_1}, \\[2mm] \rho_w \times \dfrac{V \times D_H}{\mu_w}, \dfrac{g \times \mu_w^4}{\rho_w \times \sigma^3}, \ldots \end{array} \right) \tag{3.2}$$

where d_c is the critical flow depth: $d_c = (Q^2 / (g \times B_{cell}^2))^{1/3}$, V_c is the critical flow velocity, Q is the water discharge, and D_H is the equivalent pipe diameter, or hydraulic diameter.

In Equation (3.2), in the term of the right side, the seventh term is the inflow Froude number Fr_1, while the eighth and ninth terms are the Reynolds number Re and Morton number Mo, respectively. Note that the Morton number is introduced because it becomes a constant in most hydraulic model studies, when air and water are used in both laboratory experiments and prototype flows (Kobus, 1984; Chanson, 2009b).

Traditionally hydraulic laboratory modeling is conducted using geometrically similar models (Chanson, 1999b, 2004). Geometric similarity implies that the ratios of prototype characteristic lengths to model lengths are equal. If any form of similarity (i.e. geometric, kinematic, or dynamic) is not satisfied, scale effects may take place, yielding substantial differences between the laboratory data extrapolation and the culvert prototype structure performances.

In a physical model, true similarity can be achieved if and only if all dimensionless parameters or $\prod$ terms have the same values in both laboratory and full-scale structure:

$$
\begin{aligned}
\mathrm{Fr_m} &= \mathrm{Fr_p} \\
\mathrm{Re_m} &= \mathrm{Re_p} \\
\mathrm{Mo_m} &= \mathrm{Mo_p}
\end{aligned}
\tag{3.3}
$$

where the subscripts m and p refer to the laboratory model and full-scale conditions, respectively.

Open channel flows, including culvert flows, are traditionally investigated based upon a Froude similarity because gravity effects are important (Henderson, 1966; Novak and Cabelka, 1981). When the same fluids – air and water – are used in the laboratory and at full scale, the Froude and Morton similarities are applied simultaneously. Then the Reynolds number may be grossly underestimated in small-size laboratory flumes.

A dimensional analysis must be similarly undertaken for fish motion in a turbulent culvert barrel flow (Alexander, 1982; Wang and Chanson, 2018a). Considering the upstream passage of a fish in a prismatic box culvert barrel with steady turbulent flow conditions, a dimensional analysis gives a series of relationships between the fish motion at a given location (x, y, z), fish properties (including species), the channel boundary conditions, turbulent flow properties, fluid properties, and physical constants:

$$
\vec{U}, u', O_2, \tau_f, \ldots = F_3 \begin{pmatrix} x, y, z, \\ L_f, l_f, h_f, \rho_f, \text{specie}, \\ B, k_s, \theta, \\ d, V, v', L_t, T_t, \\ \rho_w, \mu_w, \sigma, g, \ldots \end{pmatrix}
\tag{3.4}
$$

where U is the fish speed positive upstream; u' is a fish speed fluctuation; O_2 is the oxygen consumption; τ_f is the fish response time; L_f, l_f, and h_f are, respectively, the fish length, thickness, and height; and ρ_f is the fish density.

Note that Equation (3.4) ignores the effects of fish fatigue, heat transfer, and metabolism. The application of the $\prod$-Buckingham theorem shows that Equation (3.4) may be transformed into a dimensionless form:

$$
\frac{U}{V_c}, \frac{u'}{v'}, O_2, \frac{\tau_f}{T_t}, \ldots = F_4 \begin{pmatrix} \dfrac{x}{d_c}, \dfrac{y}{d_c}, \dfrac{z}{d_c}, \\[2mm] \dfrac{L_f}{L_t}, \dfrac{l_f}{L_t}, \dfrac{h_f}{L_t}, \dfrac{\rho_f}{\rho_w}, \text{specie}, \\[2mm] \dfrac{B}{d_c}, \dfrac{k_s}{d_c}, \theta, \\[2mm] \mathrm{Fr}, \mathrm{Re}, \mathrm{Mo}, \dfrac{L_t}{d_c}, T_t \times \sqrt{\dfrac{g}{d_c}}, \ldots \end{pmatrix}
\tag{3.5}
$$

The finding, that is, Equation (3.5), demonstrates several key dimensionless variables most relevant to the upstream passage of fish in a turbulent culvert barrel flow. These fundamental parameters encompass the ratio u'/v' of fish speed fluctuations to fluid velocity fluctuations; the ratio τ_f/T_t of fish response time to turbulent time scales; the ratios of fish dimension to turbulent length scale, L_f/L_t, and h_f/L_t; and the fish species (Wang and Chanson, 2018a). Considering the upstream fish passage in a turbulent culvert flow, the extrapolation of the laboratory model data to the full-scale culvert will be achievable only if all the relevant key dimensionless parameters shown in Equation (3.5) are the same in the laboratory and in the full-scale culvert structure.

A few studies recorded quantitative detailed characteristics of both fish motion and fluid flow (Nikora *et al.*, 2003; Plew *et al.*, 2007; Wang *et al.*, 2016a). Fewer investigations reported fish speed fluctuations and fluid velocity fluctuations, as well as fish response time and integral time scales (Wang *et al.*, 2016a; Cabonce *et al.*, 2018). All the results showed that a number of key parameters, including the ratios u'/v', τ_f/T_t, and L_f/L_t, are scale dependent when the same fish are used in laboratory and in the field, as shown by Equation (3.5). Basically, a complete similarity between laboratory data and full-scale observations may be unachievable. One must seek either an incomplete similitude, some approximate estimate, or an alternative approach. The latter may be some full-scale testing under carefully controlled flow conditions, as discussed in Appendix C (Section C.3).

Commentary on physical modeling of fish passage structures

by Professor Daniel B. Bung, FH Aachen University of Applied Sciences, Germany

Physical modeling is a classical tool in hydraulic engineering to analyze complex flows. Although flood analyses of large areas are mostly investigated by means of (mostly) depth-averaged numerical models nowadays, detailed investigations of three-dimensional fluid–structure interactions are still subject to this classical approach. Particularly, when a second phase comes into play, that is, the so-called two-phase flow problems, numerical methods (computational fluid dynamics [CFD]) still require certain model assumptions to address phase interactions. While the complexity is yet high enough in the case of, for example, sediments or air as a second phase, it becomes even higher when the water phase "interacts" with fish, having a species-dependent, particular behavior that is difficult to generalize or even to quantify.

When dealing with physical modeling, researchers and engineers need to carefully select a feasible model scale in order to avoid so-called scale effects, which in general are caused by improperly accounting for turbulence in the model (i.e. turbulence is underestimated). It was shown in this chapter that knowledge of turbulence characteristics within a culvert is essential for evaluating its fish passage potential and thus requires large-scale physical models to adequately reproduce the flow field in the prototype. However, because fish are not scalable and smaller, juvenile fish may behave differently from their adult counterparts, large-scale models alone are mostly insufficient, but additional protype-scale model tests may become essential to draw any final conclusions. This issue becomes more problematic for larger fish species, which require larger, more expensive model setups. Additionally, whenever possible, fish passage studies should include prototype studies (i.e. in its natural environment) to avoid having fish behavior affected by the laboratory conditions – it must be noted that (a) fish may get trained after a first test, knowing the optimum path for passage in subsequent tests, (b) fish migration motivation may depend on the time of day, (c) water temperature and oxygen content may affect the fish's swimming potential, etc. It is difficult to account for all these factors with classical tools, such as the dimensional analysis proposed in this chapter. However, these tools are established and approved physics-based methods. Shortcomings from the imprecise quantification of fish behavior need to be compensated for by collecting enough data with a statistically sufficient database.

The idea of observing fish passage behavior in laboratory environments is not new. Yet despite all the mentioned difficulties, supported by the continuous development of new instrumentation and data processing methods, it is an important and the most reliable approach to address this important problem. Many water bodies have been heavily modified in the last century, and the ecological continuity has been interrupted by weirs and dams all over the globe. Restoring this continuity is one of the most crucial tasks hydraulic engineers and researchers, together with biologists, are dealing with today.

3.3   Fish behavior and kinematics in a box culvert barrel

Open channels and culvert barrel flows are modeled based upon a Froude similarity because gravity effects are important in terms of the hydrodynamics (Henderson, 1966). Viscous-scale effects are likely to be experienced in very small-size models, water tunnels, and water tubes, and the results cannot be extrapolated to a full-scale culvert without major bias (i.e. scale effects). When the hydrodynamics and fish kinematics are considered together, the similitude requirements become impossible to fulfill unless full-scale studies are undertaken (Section 3.2). Only measurements in a real culvert and full-scale laboratory experiments are appropriate. A review of fish swimming performance tests conducted in a near-full-scale facility is discussed herein (Table 3.1). The channel was 12 m long and 0.5 m wide, corresponding to a typical culvert barrel cell beneath a two-lane road embankment. The experimental flow conditions are summarized in Table 3.1, including the boundary conditions, water discharge, bulk velocity, fish sample size, fish data, and water temperature. During each series of tests, the flow rate was kept constant, irrespective of the boundary treatment. The methodology delivered biological data compatible to engineering design procedures and applicable by professional engineers (Chanson, 2019), as well as enabled a direct comparison of fish swimming performances and behavior in culvert barrel between different types of boundary treatments.

Typical fish endurance test results are presented in Figure 3.4, regrouping the cumulative percentages of fish swimming after test durations ranging from 1 to up to 20 minutes for two different series of tests with different discharges, boundary conditions, and fish species (Table 3.1). For the larger discharge, a sizable number of fish fatigued before the end of testing – 12 out of 20 – in the smooth boundary flume (Fig. 3.2A). The fish position observations indicated that the fish swam against the current, mostly next to the bottom corners and along the sidewall of the channel (Fig. 3.3) (Appendix C). All the results indicated that the fish swam in the bottom corners and along the sidewalls for more than 90%

Table 3.1 Review of laboratory studies on fish swimming in a 12 m long and 0.5 m wide culvert barrel model: fish data (mass m_f and total length L_t)

Reference	Q	d	Vmean	T	Fish species	Nb of fish	Fish mass m_f (g)	Fish length L_t (mm)
	(m^3/s)	(') (m)	(') (m/s)	(°C)				
(1)	(2)	(3)	(4)	(5)	(6)	(7)	(8)	(9)
Wang et al. (2016a)								
Smooth channel	0.0261	0.123	0.424	24.5	Duboulay's rainbowfish (Melanotaenia duboulayi)	22	2.75 ± 0.65	68.5 ± 6.3
Rough bed and smooth sidewalls	0.0261	0.133	0.392	±0.5	Duboulay's rainbowfish (Melanotaenia duboulayi)	23	3.6 ± 1.08	74.0 ± 5.5
Rough bed and rough left sidewall	0.0261	0.129	0.424		Juvenile silver perch (Bidyanus bidyanus)	23	39.7 ± 33.7	145 ± 31.5
					Duboulay's rainbowfish (Melanotaenia duboulayi)	23	3.2 ± 1.07	70.5 ± 8.0
Cabonce et al. (2018, 2019)								
Smooth channel	0.0556	0.162	0.686	24.5	Juvenile silver perch (Bidyanus bidyanus)	20	1.50 ± 1.16	53.0 ± 11.8
Corner baffles (h_b = 0.067 m)	0.0556	0.1625	0.684	±0.5		26	1.30 ± 0.85	47.0 ± 9.6
Corner baffles (h_b = 0.133 m)	0.0556	0.173	0.643			26	3.70 ± 2.81	70.5 ± 16.7
Corner baffles (h_b = 0.133 m) with perforation (Ø = 13 mm)	0.0556	0.173	0.643			15	3.20 ± 1.40	66.0 ± 8.7

Notes: d: water depth; h_b: isosceles triangular baffle size; Q: water discharge; T: water temperature; V_{mean}: bulk velocity; fish data: median value ± standard deviation; ': values recorded 8 m downstream of the flume's entrance.

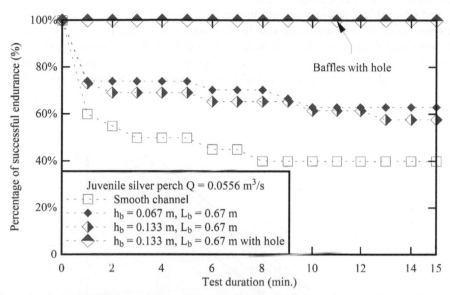

(A) Juvenile silver perch (*Bidyanus bidyanus*) in smooth flume without and with small triangular corner baffles along the left bottom corner for a subcritical flow: Q = 0.0556 m³/s, θ = 0

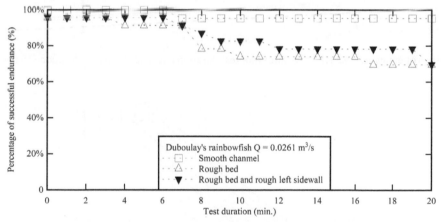

(B) Duboulay's rainbowfish (*Melanotaenia duboulayi*) in smooth and rough channels for a relatively small flow: Q = 0.0261 m³/s, θ = 0

Figure 3.4 Cumulative endurance test duration data for small-body-mass fish negotiating upstream passage in a 12 m long and 0.5 m wide culvert barrel flume for less-than-design discharges (Table 3.1) – comparison between different boundary treatments

of the time. The findings were consistent with a number of studies with other small-bodied and larger fish species (Gardner, 2006; Jensen, 2014; Duguay *et al.*, 2018; Goodrich *et al.*, 2018). Visual observations, fish trajectory data, and speed time series indicated that for both fish species the time series could be subdivided into (a) quasi-stationary motion where fish speed fluctuations were small, (b) short upstream motion facilitated by a few strong tailbeats, and (c) burst swimming when the fish would rapidly cross the observation window. See, for example, the movies CIMG1655.mov, CIMG2672.mov, CIMG1647.mov, CIMG2725. mov, and CIMG1651.mov in a smooth box culvert channel (Appendix F). The most common observation of fish swimming was the first one: that is, quasi-stationary motion with small fish speed fluctuations. Altogether, the fish took from about 1 minute to more than 20 minutes to swim the entire culvert flume, spending most of their time in the bottom corners and along the sidewalls. Typical individual fish trajectories are shown in Figure 3.5. In one case (Fig. 3.5A), the fish traversed almost half the barrel length in 15 minutes, which was the maximum duration of that test.

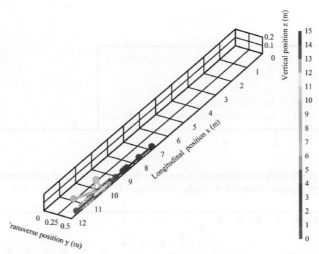

(A) Fish properties: m_f = 4.5 g, L_f = 73 mm, swim test duration until maximum test duration (15 minutes)

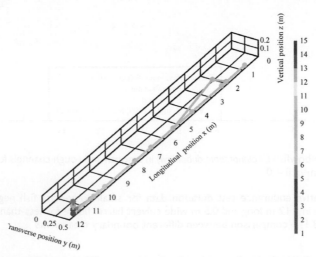

(B) Fish properties: m_f = 0.6 g, L_f = 37 mm, swim test duration until complete culvert barrel traverse time (11 minutes)

Figure 3.5 Typical fish trajectories during fish testing: Q = 0.0556 m³/s, d = 0.161 m, V_{mean} = 0.69 m/s, θ = 0, flow direction of top right to bottom left, juvenile silver perch (*Bidyanus bidyanus*) – the scale in minutes corresponds to the maximum test duration (15 minutes) or the culvert barrel traverse time

Fish kinematics, based upon high-speed video movies (Appendix F), deliver seminal insights into fish behavior and trajectories, fish speed and acceleration, tailbeat frequencies, and fish swimming energetics. The most interesting, and probably the most important, investigations concern the simultaneous characterization of both fish and fluid kinematics (Nikora *et al.*, 2003; Plew *et al.*, 2007; Wang *et al.*, 2016a; Cabonce *et al.*, 2018). Experimental data are reviewed in Appendix C, and high-speed movies are presented in Appendix F. Figure 3.6 presents the typical probability density functions (PDFs) of fish speed and acceleration. The fish swimming speed variability may be compared to the distribution of a longitudinal fluid velocity component at the location where the fish is tracked. For small-body-mass fish species, the ratio of fish speed to fluid velocity standard deviations was typically within $0.1 < u_x'/v_x' < 1$ with a median value about 0.25, independent of the fish species; total length and mass, u_x' being the fish speed fluctuations; and v_x' being the velocity fluctuations at the observed fish location (Fig. 3.7A). In Figure 3.7A, the data are presented as a function of the dimensionless fish length L_f/k_s, with L_f the total fish length and k_s the equivalent sand roughness height of the channel. Physically, the roughness height k_s characterizes the rugosity of the channel boundaries and is related to the size of vortical structures in the vicinity of the wall (Hong *et al.*, 2011). Basically the ratio L_f/k_s provides some indication of the ratio of fish length to turbulence scale. The key finding, that is, $0.1 < u_x'/v_x' < 1$, may suggest that swimming in the channel corner may allow fish to minimize the energetic costs associated with changes in acceleration (Nikora *et al.*, 2003; Wang and Chanson, 2018a, 2018b).

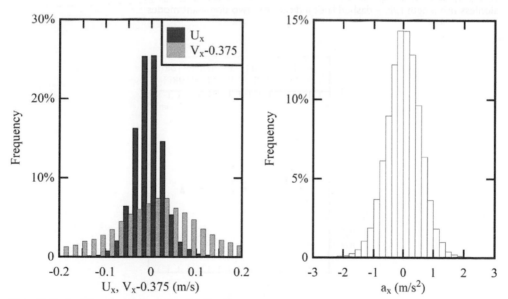

(A, Left) Probability density function (PDF) of longitudinal fish speed U_x – comparison with fluid velocity data V_x at the fish tracking location

(B, Right) PDF of longitudinal fish acceleration a_x

Figure 3.6 Fish kinematics of a small-bodied fish swimming along the rough sidewall of an asymmetrically roughened culvert barrel channel – Duboulay's rainbowfish (*Melanotaenia duboulayi*): m_f = 3.6 g, L_f = 82 mm – Q = 0.0261 m³/s, d = 0.129 m, V_{mean} = 0.423 m/s

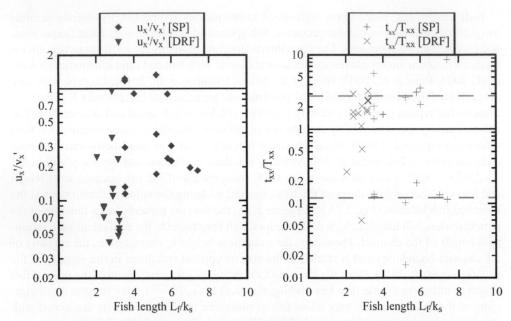

(A, Left) Ratio of fish speed to fluid velocity standard deviation u_x'/v_x' as a function of the dimensionless fish length L_f/k_s

(B, Right) Ratio of fish speed to fluid velocity auto-correlation time scales as a function of the dimensionless fish length L_f/k_s – dashed lines indicate the two dominant modes

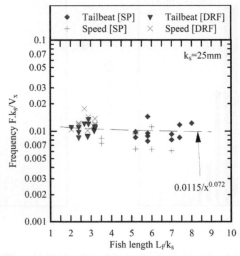

(C) Dimensionless frequencies of fish tailbeat and characteristic fish speed fluctuations $F \times k_s / V_x$ as functions of the dimensionless fish length L_f/k_s

Figure 3.7 Fish kinematic properties for juvenile silver perch (*Bidyanus bidyanus*) [SP] and adult Duboulay's rainbowfish (*Melanotaenia duboulayi*) [DRF] – flow conditions: Q = 0.0261 m³/s, d = 0.129 m, V_{mean} = 0.423 m/s, asymmetrically roughened culvert barrel channel – comparison with fluid velocity properties at the observation location

As shown earlier, another key parameter is the ratio t_{xx}/T_{xx} of fish speed to fluid velocity auto-correlation time scales. The fish speed auto-correlation time scale gives some information on the reaction time of the fish, while the fluid velocity auto-correlation time scale is a rough measure of the longest connection in the turbulent behavior (O'Neill et al., 2004; Chanson, 2009b). The ratio t_{xx}/T_{xx} basically characterizes the fish response time relative to the characteristic time scale of large turbulent structures. Experimental observations yielded $0.03 < t_{xx}/T_{xx} < 5$ with a median value about 1.5 (Fig. 3.7B), suggesting that the small-bodied fish tend to react predominantly to the larger vortical structures and did not modulate their speed in response to small and short-lived turbulent structures. Note the data set hinting a bimodal distribution: $t_{xx}/T_{xx} \sim 0.12$ and $t_{xx}/T_{xx} \sim 2.8$, highlighted in Figure 3.7B (horizontal dashed lines). The result could hint for two preferential responses of fish to turbulence and turbulent structures. In one mode ($t_{xx}/T_{xx} < 1$), the fish would react passively to vortical structures, with their slow response possibly enabling them to be advected by the flow turbulence (e.g. in recirculation zones and secondary currents). In the second mode ($t_{xx}/T_{xx} > 1$), the fish would be proactive and respond very rapidly to a change in turbulent flow conditions, using proactively the changes in instantaneous flow conditions to migrate upstream.

A further fish swimming characteristic is the fish tailbeat frequency (Fig. 3.7C). Observations of small-body-mass fish showed that the tailbeat frequency spanned over a narrow range under the tested conditions: $F \times k_s / V_x \sim 10^{-2}$ on average. In Figure 3.7C, the data presented some correlation in terms of the dimensionless fish length:

$$\frac{F \times k_s}{V_x} = 0.0115 \times \frac{L_f}{k_s} \tag{3.6}$$

The tailbeat frequency data may be further compared to the characteristic frequency of the longitudinal fish speed. An example of the latter is illustrated in Figure 3.7C, showing that the characteristic fish speed frequency may be used as a proxy of the tailbeat frequency.

3.4   Discussion

Detailed full-scale laboratory data highlighted a number of issues that deserve some discussion (Table 3.1). The fish speed fluctuations were systematically smaller than the turbulent velocity fluctuations at the fish location (i.e. $u_x' < v_x'$). In turn, the fish accelerations were small and the corresponding inertial force was minimal. The fish swimming accelerations have some important implication in terms of energy expenditure required to swim against the current over a period of time (Plew et al., 2007). Power is required to overcome friction and form drag (Videler, 1993), and additional energy is spent during the acceleration phases. The combined power to overcome skin friction and form drag is proportional to the cube of fish speed relative to the mean fluid motion that is, power $\propto (U_x + V_x)^3$, while the power required during acceleration periods is basically the fish mass times acceleration times relative fish speed: power $\propto m_f \times a_x \times (U_x + V_x)$ (Wang and Chanson, 2018a). A minimization of the fish accelerations would yield to smaller inertial forces and lesser energy consumption.

In the view of ecologically friendly engineering design, which has motivated many studies, a comprehensive fish behavior study would be beneficial, including how fish sense fluid flow turbulence to select the optimum upstream path in turbulent flows. For example, many observations reported that fish swim preferentially next to the bottom corners in rectangular channels: that is, in regions of low fluid velocity, but very high turbulence and intense secondary currents. Further investigations could consider the characteristic fish acceleration frequencies, as well as the auto-correlation time scales of the fish acceleration and a quantitative description of fish energy consumption during upstream migration.

Note

1 It is also called the Vaschy-Buckingham theorem after the French engineer Aimé Vaschy (1857–1899) and American physicist Edgar Buckingham (1867–1940).

Chapter 4

Hydraulic engineering design for upstream fish passage in standard box culverts: general concepts and design guidelines

4.1 Presentation

Based upon current engineering design practices, the hydraulic characteristics of the structure are the design discharge and the maximum acceptable afflux at design flow conditions. An important hydrodynamic feature is whether the barrel runs full or not (Fig. 4.1). Most culverts are designed to operate as open channel systems up to the design flow conditions, often with critical flow conditions occurring in the barrel in order to maximize the discharge per unit width and to reduce the barrel cross-section during the design event. For standard culverts, the current engineering design procedure can be divided into two parts. First a system analysis leads to the selection of (a) the design rainfall and runoff event, yielding an

(A) Masonry box culvert inlet along Le Rat river at Le Temple (22), France, on 28 June 2019 during low-flow condition – 19th-century construction

Figure 4.1 Multicell standard box culvert operation

(B) Concrete box culvert inlet along Cubberla creek beneath Goolman Street, Chapel Hill QLD, Australia, on 30 March 2017 at less-than-design flow

Figure 4.1 (Continued)

estimate of the design discharge Q_{des}, and (b) the maximum acceptable afflux at design flow conditions. In the second stage, the barrel size is selected by an iterative procedure, in which both inlet control and outlet control calculations are conducted.[1] At the end, the optimum hydraulic size is the smallest barrel size allowing for inlet control operation at design flow conditions.

In this chapter, a new design methodology is developed to assist with the upstream passage of weak-swimming fish in standard box culverts. The proposed design approach is primarily aimed at the design of standard box culverts, focusing on the culvert barrel and assuming a structure built in a flood plain with a mild slope. Designs of culverts outside this scope will not be detailed.

Commentary: what constitutes effective fish passage through remediated structures?

by Dr. Cindy Baker, NIWA, New Zealand

In New Zealand around 72% of our freshwater fish species are threatened or in decline. Aside from the degradation of adult habitats, one of the most significant causes of their decline is the construction of in-stream structures such as culverts and weirs that prevent fish from accessing critical life-stage habitats. In September 2019, new government policies were proposed to address fish passage management across New Zealand. The draft national policy statement proposes regional authorities must identify existing in-stream structures within their region, evaluate the risk to fish passage, prioritize and remediate those structures impeding fish passage, and monitor and evaluate the performance of remediated structures. However, implementation of the proposed policy will require two key knowledge gaps to be addressed:

- As poorly positioned culverts form one of the main fish passage migration barriers in New Zealand, there is an urgent need to develop effective solutions for passing native freshwater fish species through culvert barrels.
- In assessing the efficacy and performance of remediated structures, how much passage is enough?

For any remediated structure, the performance of the solution will vary between and within sites. Focusing on culvert passage, some of the key factors influencing a solution's efficacy are changing flow, where water depths and water velocities will vary, the width and gradient of the stream relative to the culvert width, gradient and length, and the position of the culvert in the catchment. With respect to the culvert location, the distance inland and elevation will dictate both the target species requiring passage and the abundance of migrants reaching the structure. To maintain fish communities above the structure, the size of the upstream catchment will also influence the required efficacy of the retrofitted structure. Although achieving 100% passage success for all species at all times is desirable, it is not always realistic when retrofitting existing structures and may not be necessary for every situation. For example, achieving 100% passage success at 80% of the flows experienced during the migration season may be sufficient to maintain upstream fish populations. In New Zealand, there is currently not enough monitoring of remediated culverts to determine what constitutes passage success relative to culvert location, type, solution implemented, and flow regime. Therefore, enhancing the toolbox of fish passage solutions, along with monitoring of solutions, will be crucial in developing a framework to enable regional authorities to evaluate performance success of retrofitted structures and ensure they meet upcoming policy requirements.

4.2 Basic concepts

With current hydraulic engineering design practices, the optimum size of a culvert is the smallest barrel size allowing for inlet control operation (Chapter 2). The approach is focused on design flow conditions solely and rarely considers less-than-design flow conditions, that is, $Q < Q_{des}$, although fish passage may occur as soon as the water discharge is nonzero: $Q > 0$. New design guidelines for fish-friendly box culverts are needed. A practical challenge is matching biological data (e.g. swimming performances) to engineering requirements and hydrodynamic measurements because of a lack of standardization in swim tests (Kemp, 2012; Katopodis and Gervais, 2016).[2]

New hydraulic engineering design guidelines are considered, based upon three key concepts:

(A) The culvert design is optimized for fish passage for water discharges $Q < Q_T$, and it is optimized in terms of flood capacity for $Q_T < Q < Q_{des}$, with Q_T an upper threshold discharge for less-than-design flow with $Q_T < Q_{des}$.

(B) Upstream fish passage is facilitated by providing a sizable low-velocity zone (LVZ):

(B.1) Since fish predominantly swim upstream next to the channel corners and sidewalls, including small-bodied fish species (Chapter 3), the swimming performance data are related to a fraction (i.e. percentage) of the wetted flow area where:

$$0 < V_x < U_{fish} \tag{4.1}$$

with V_x the local time-averaged longitudinal velocity component and U_{fish} a characteristic swimming speed of targeted fish species. A truly novel aspect of this approach is the provision of a minimum relative flow area where the longitudinal water velocity is less than a characteristic fish swimming speed (Fig. 4.2).

(B.2) The LVZ width and depth in the bottom corners must encompass the size of the targeted fish species.

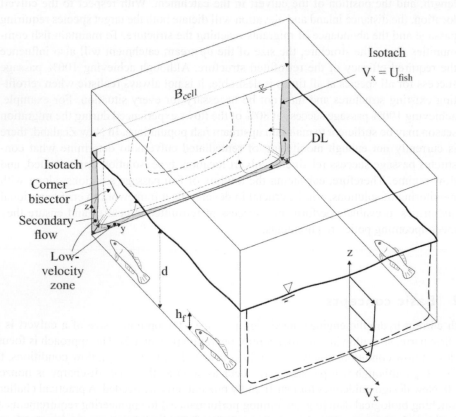

Figure 4.2 Sketch of upstream passage of fish in the low-velocity zones of a smooth box culvert barrel, with flow direction from top left to bottom right

(C) The culvert invert and barrel walls are smooth and upstream fish passage is provided without any other form of boundary treatment and appurtenance.

One may contrast past and novel design guidelines of fish-friendly culverts (Table 1.1). In the past, fish-friendly culverts have been designed with respect to bulk velocity criteria, that is, a maximum bulk velocity largely attributed to the culvert across the full flow range, neither of which appear to be relevant to reality. In contrast, the new design guidelines consider the fact that fish predominantly swim along the sidewalls and at higher discharges, fish passage is generally not possible given the constraints of the culvert design relative to the physical capabilities of small-bodied fish.

4.3   Implementation of basic concepts

The design method aims to be sound, simple, economically acceptable, and meet engineering standards. Three practical questions must be discussed with respect to the design parameters influencing the size and cost of standard box culverts: (a) what is the effect of the relative threshold Q_T/Q_{des}, (b) what is the influence of the percentage of low-velocity area, and (c) what is the impact of the characteristic fish swimming speed U_{fish}? A sensitivity analysis was recently conducted regarding these queries (Leng et al., 2019). The basic findings demonstrated unequivocally that the cost of a fish-friendly box culvert increases with decreasing characteristic fish speed U_{fish}, increasing discharge threshold Q_T/Q_{des}, and increasing percentage of low-flow area:

$$\text{Culvert Cost} \uparrow \equiv \begin{cases} U_{fish} \downarrow \\ Q_T / Q_{des} \uparrow \\ \%\text{flow area} \uparrow \end{cases} \tag{4.2}$$

Based upon a combination of engineering calculations, detailed physical modeling for several flow boundary conditions during which fish endurance and behavior were tested (Appendix C), and discussions between biologists, scientists, and hydraulic engineers, the following guidelines are recommended:

- $\dfrac{Q_T}{Q_{des}} < 0.10 \text{ to } 0.3$ (4.3a)

- U_{fish} set by fisheries department based upon fish swimming performance data for catchment fish species (4.3b)
- Fifteen percent of flow area where $V_x < U_{fish}$ for $Q < Q_T$ (4.3c)
- Bottom corner LVZ size $> h_f$ (4.3d)

where h_f is the size of the targeted fish species.

4.3.1   Discussion

The present approach delivers a physically based rationale for fish-friendly standard box culvert design, embedding state-of-the-art aerodynamic and hydrodynamic calculations into current hydraulic engineering design methods to yield cost-effective design outcomes. By bridging the gap between engineering and biology, such an approach is innovative in an effort to restore catchment connectivity. The method is general, yet simple and cost-effective enough to be widely endorsed by the various stakeholders.

The proposed concepts for fish-friendly box culvert design were developed in particular for fish species with weak-swimming abilities[3] and for which excessive barrel velocities are too often a barrier to upstream fish passage. However, the present approach may be applied to other fish species, because it is physically based, with deterministic scientific considerations. LVZs are provided along the wetted perimeter, that is, next to the culvert barrel corners and sidewalls, where fish prefer to swim (Fig. 4.2) and where they can minimize their energy expenditure. See the movies CIMG1655.mov, CIMG2672.mov, CIMG1647.mov, CIMG2725.mov, and CIMG1651.mov in Appendix F for examples. The current approach relies upon an accurate, physically based knowledge of the entire velocity field in the culvert barrel.

The influence of the relative threshold Q_T/Q_{des}, critical fish speed U_{fish}, and percentage of flow area on the size of box culvert structures was specifically discussed by Leng *et al.* (2019). For a smooth culvert barrel invert at natural ground level, the results showed that the increase in culvert size and hence cost become very significant for $U_{fish} < 0.3$ m/s and $Q_T/Q_{des} > 0.3$, when providing 15% flow area with $0 < V_x < U_{fish}$ in a smooth barrel. When the characteristic swimming speed of the targeted fish species is less than 0.3 m/s ($U_{fish} < 0.3$ m/s), a different design approach might be required (Chapter 6, Appendix E). An option could involve a design incorporating a barrel cell with enhanced LVZs, for example, boundary roughening and adding appurtenance such as baffles. Another option might be to consider a bridge structure instead.

In terms of hydraulic engineering calculations, the fundamental concepts (Section 4.2) lead to a two-stage design. First the minimum number of cells $(N_{cell})_{des}$ is calculated to achieve inlet control at design flow conditions, based upon current standards for optimum flood capacity design at the culvert site (Chapter 2). Considerations for upstream fish passage are next embedded into the design methodology using biological considerations. When the revised fish-friendly culvert design includes a larger number of barrel cells than the original design, the afflux at the design flow conditions would be smaller than the maximum acceptable afflux. The reduction in upstream flooding might contribute to a lesser total cost of the structure: for example, with a lower embankment and reduced impact on upstream catchment. The savings might contribute to offset partially the increased cost caused by the larger number of culvert barrel cells.

4.4   Design methodology

For the hydraulic design of fish-friendly culverts, the previous approach must be expanded. The design procedure takes place in two stages:

1   Finding the optimum design to pass the design flow discharge Q_{des} for a standard box culvert
2   Checking whether additional culvert cells are required or not required to pass fish at less-than-design flow discharge (i.e. for $Q \leq Q_T$)

Calculations are first conducted for design flow conditions. The optimum size is the smallest barrel size allowing for an afflux less than the maximum acceptable afflux at design discharge Q_{des} (Herr and Bossy, 1965; Chanson, 2004) (Chapter 2). The second part of the design corresponds to a culvert operation for less-than-design flow (i.e. $Q \leq Q_T$) for which typically 15% of the flow area must experience local time-averaged velocities less than the characteristic fish swimming speed (U_{fish}) (Equation (4.3c)). When choosing the minimum fish swimming speed for the site in relation to the species and size classes of expected fish, the local fisheries department should be consulted.

Both stages of calculations are iterative processes. The optimum design for Q_{des} with a maximum acceptable afflux h_{max} would basically yield the minimum number (N_{cell})$_{des}$ of boxes required to achieve inlet control for a multicell culvert (Chapter 2). Two alternative methods may be used, being an engineering design nomograph (Concrete Pipe Association of Australasia, 2012) or a set of theoretical equations based on critical conditions and con-servation of energy (Henderson, 1966; Chanson, 2004). The theoretical equations give the design discharge per unit width for inlet control:

$$\frac{Q_{des}}{B_{min}} = C_D \times \frac{2}{3} \times \sqrt{\frac{2}{3} \times g} \times (H_{hw} - z_{hw})^{3/2} \qquad \text{for } \frac{H_{hw} - z_{hw}}{D_{cell}} < 1.2 \qquad (4.4a)$$

$$\frac{Q_{des}}{B_{min}} = C \times D_{cell} \times \sqrt{2 \times g \times (H_{hw} - z_{hw} - C \times D_{cell})} \qquad \text{for } \frac{H_{hw} - z_{hw}}{D_{cell}} > 1.2 \qquad (4.4b)$$

where B_{min} is the internal barrel width, D_{cell} is the internal barrel height, ($H_{hw} - z_{hw}$) is the headwater-specific energy, often assumed to be about the headwater level: $H_{hw} - z_{hw} \approx d_{tw} +$ afflux. When the headwater is less than 1.2 times the internal barrel height, a free-surface inlet is observed and Equation (4.4a) should be used. Otherwise, a submerged entrance and free-surface barrel flow occur, and Equation (4.4b) is to be used. The constants C_D and C correspond to the shape of the inlet. For the most common square-edged inlet, $C_D = 0.9$ and $C = 0.6$.

Next, the design must be checked against the outlet control situation, using the outlet con-trol nomograph (Concrete Pipe Association of Australasia, 2012). The nomograph will give the afflux for the calculated number of cells at the outlet control. Ultimately, whichever afflux is bigger controls the flow, for example, if afflux for inlet control is greater than afflux for outlet control, inlet control is achieved; otherwise, the number of cells must be increased, and outlet control calculations are repeated until the afflux is less than h_{max} (Chapter 2, Section 2.2).

Once the number of cells (N_{cell})$_{des}$ is obtained at design flow conditions, hydrodynamic cal-culations must be performed for a single cell to examine the complete velocity field throughout the culvert cell for less-than-design flows ($Q \leq Q_T$).[4] While a box culvert barrel is a relatively simple geometry channel, a complete knowledge of the fluid flow requires advanced calcula-tions based upon a thorough characterization of the boundary conditions, including the water flow in the inlet and outlet sections, as well as in the downstream and upstream flood plains. The calculations of the entire velocity field in a barrel are complicated at less-than-design discharges. Complete numerical computational fluid dynamics (CFD) computations, even simpler one-dimensional (1D) numerical calculations, require a sound knowledge of the flood flow conditions in the natural system. The latter may be derived from gauging data or flood plain calculations (Appendix B). Physical modeling may be further considered, based upon the fundamental concepts and principles of similitude, sometimes in complement to and in support of numerical CFD computations (Leng and Chanson, 2018) (Appendix D).

Using contour plots of longitudinal velocity at different cross-sections of the culvert bar-rel, the flow area under certain velocity magnitudes can be derived. Within this second stage, a sizable portion of the flow area must be under the critical fish swimming speed U_{fish}, in line with recent hydrodynamic data (Wang and Chanson, 2018b; Cabonce et al., 2019; Zhang and Chanson, 2018). Herein the target is 15% of the wetted area (Equation (4.3c)), based upon detailed physical data. Conceptually, a different fraction could be considered.[5] An additional requirement includes a minimum LVZ size at the barrel's bottom corners to facilitate the passage of the largest targeted fish species. If the results of physical/numerical models do not satisfy the required criteria of fish-friendly design, the number of cells must be increased, and the modeling is repeated with the updated cell number configuration.

Figure 4.3 summarizes the iterative procedure for the hydraulic design of a fish-friendly standard box culvert with smooth boundaries, and Figure 4.4 presents a sketch of the LVZ delivered to assist upstream fish passage in a box culvert barrel. The readers are referred to Section 5.3.2 for an example of calculations.

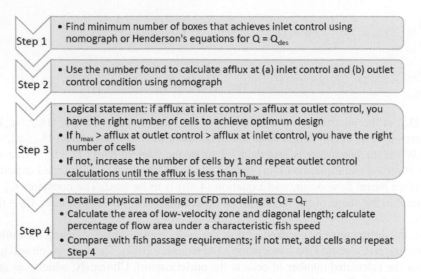

Step 1
• Find minimum number of boxes that achieves inlet control using nomograph or Henderson's equations for $Q = Q_{des}$

Step 2
• Use the number found to calculate afflux at (a) inlet control and (b) outlet control condition using nomograph

Step 3
• Logical statement: if afflux at inlet control > afflux at outlet control, you have the right number of cells to achieve optimum design
• If h_{max} > afflux at outlet control > afflux at inlet control, you have the right number of cells
• If not, increase the number of cells by 1 and repeat outlet control calculations until the afflux is less than h_{max}

Step 4
• Detailed physical modeling or CFD modeling at $Q = Q_T$
• Calculate the area of low-velocity zone and diagonal length; calculate percentage of flow area under a characteristic fish speed
• Compare with fish passage requirements; if not met, add cells and repeat Step 4

Figure 4.3 Flow chart of the methodology for the optimum design at fish-friendly box culverts

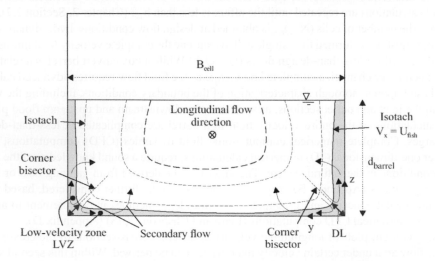

Figure 4.4 Sketch of low-velocity zone (shaded area) in a box culvert, including definition of 45 degrees diagonal (i.e. corner bisector) from bottom corner of box culvert, looking downstream

4.5 Simplified design guidelines

4.5.1 Presentation

The properties of a culvert operating at less-than-design flow may be derived from hydraulic engineering calculations (Appendix B). For a mild flood plain slope, the culvert operates with outlet control for $Q < Q_{des}$. The flow in the entire culvert system is subcritical, and the calculations are best started from the downstream end (i.e. the tailwater conditions). The complete calculations involve (a) the estimate of the form losses in the culvert outlet and transition to the downstream flood plain, (b) the hydrodynamic calculations of the culvert barrel flow and the boundary friction, and (c) the application of the Bernoulli principle to the flow convergence in the transition from the upstream flood plain to the inlet and in the culvert inlet. For a culvert in a steep catchment, the same type of hydrodynamic calculations is conducted, albeit starting from the upstream end (i.e. the headwater conditions).

When fish passage is a requirement for $Q < Q_T$, a number of basic assumptions may be considered to simplify the hydraulic engineering design calculations:

1 The flood plain's longitudinal slope is mild and the flood plain operates with subcritical flow conditions for $Q < Q_T$.
2 The free-surface elevation in the barrel equals the tailwater free-surface elevation $Q < Q_T$; in the first approximation, the water depth d_{barrel} in the barrel is equal to the tailwater depth d_{tw}:

$$d_{barrel} \approx d_{tw} \tag{4.5}$$

   for a culvert structure with invert set at ground level.
3 There is a monotonic relationship between the relative LVZ area, where the dimensionless velocity V_x/V_{mean} is less than U_{fish}/V_{mean}, and the dimensionless targeted fish swimming speed U_{fish}/V_{mean}, with V_{mean} the bulk velocity in the culvert barrel and U_{fish} the characteristic fish swimming speed.
4 The length DL of 45 degrees diagonal from the bottom corner (Fig. 4.4) is a function of the targeted velocity $V_x = U_{fish}$, bulk velocity V_{mean}, water depth d, and channel width B only:

$$DL = F_1(V_x, V_{mean}, d, B) \tag{4.6}$$

Practically, the most stringent hydrodynamic conditions for upstream passage of small-bodied fish take place for $Q = Q_T$. In turn, the hydraulic design calculations for upstream fish passage are typically focused on $Q = Q_T$. In the next paragraphs, each basic assumption is discussed and its practical implementation is developed.

4.5.2 Mild flood plain

In hydraulic engineering, a channel slope is termed "mild" when the uniform equilibrium flow depth is larger than the critical flow depth[6] and the uniform equilibrium flow is subcritical (Chow, 1959; Henderson, 1966). Such a definition implies physically that the concept of a mild slope is a function of both the bed slope and the flow resistance: that is, of the flow rate and channel roughness height (Chanson, 2004, p. 95). Here it is assumed that the flood plain longitudinal slope is mild and the flood plain operates with subcritical flow conditions for $Q < Q_T$.

Subcritical open channel flows are best controlled from downstream (Henderson, 1966; Chanson, 2004). Numerical calculations of steady subcritical flows should thus start from the downstream end of the reach, that is, the flood plain downstream of the culvert system, or tailwater conditions.

For a culvert installed in a steep catchment, the hydrodynamics of the flow through the entire culvert system are much more complicated and outside the scope of this text. Engineers would need to undertake more advanced calculations complemented by detailed physical and/or numerical modeling.

4.5.3   Barrel flow depth

For a discharge substantially much smaller than the design discharge (e.g. $Q = Q_T$) the energy losses in the barrel outlet and in the culvert barrel are relatively small. The application of the continuity and Bernoulli principle yields in the first approximation:

$$d_{barrel} \approx d_{tw} \tag{4.5}$$

assuming implicitly that the culvert invert is placed at the natural ground level. The application of the equation of conservation of mass in an integral form[7] gives an analytical expression for the bulk velocity in the culvert barrel:

$$V_{mean} = \frac{Q_{cell}}{B_{cell} \times d_{barrel}} \tag{4.7}$$

where Q_{cell} is the water discharge in a single cell[8] and B_{cell} is the internal cell width.

Equation (4.5) is based upon some approximation in relation to the exit loss, corresponding to $V_{mean}^2/(2 \times g) \ll d_{barrel}$. Basically, Equation (4.5) should only be considered within the earlier assumption (see item 1) of a culvert located in a mild flood plain slope for $Q \leq Q_T$.

4.5.4   Fraction of wetted area where $V_x/V_{mean} < U_{fish}/V_{mean}$

LVZs are preferred swimming zones for fish, and small-bodied fish favor swimming in the channel corners and next to the sidewalls during upstream passage (e.g. movies CIMG1655. mov, CIMG2672.mov, and CIMG1647.mov in Appendix F). Figures 4.2 and 4.4 present schematics of typical LVZs in a smooth box culvert barrel. Upstream fish passage is deemed achievable when the LVZ in the culvert barrel is sizable enough for the fish to traverse the structure.[9]

Detailed physical experiments and CFD calculations showed a monotonic relationship between the fraction of an LVZ wetted area where $V_x < U_{fish}$ and the percentage of bulk velocity $V_x/V_{mean} = U_{fish}/V_{mean}$ (Fig. 4.5). Figure 4.5 presents detailed physical and numerical CFD data obtained with less-than-design discharges ranging up to 0.3 m³/s per barrel cell and internal widths from 0.1 m to 2.4 m.

In a fully developed flow, the velocity distributions tend to follow closely a power law (Henderson, 1966; Chanson, 2004):

$$\frac{V_x}{V_{max}} = \left(\frac{z}{d}\right)^{1/N} \tag{4.8}$$

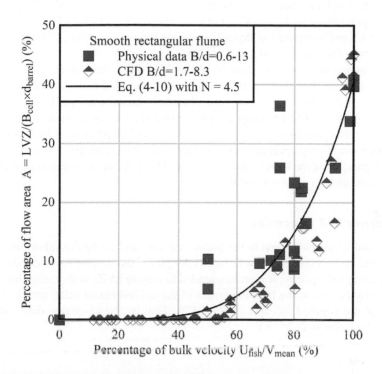

Figure 4.5 Relationship between the percentage of flow area A = LVZ/(B_{cell} × d_{barrel}) and relative longitudinal velocity U_{fish}/V_{mean} in smooth rectangular channels – comparison between Equation (6.6) assuming N = 4.5 and detailed physical experiments (Nikuradse, 1926; Nezu and Rodi, 1985; MacIntosh, 1990; Xie, 1998; Cabonce *et al.*, 2017, 2019) and CFD numerical results (Naot and Rodi, 1982; Leng and Chanson, 2018) in rectangular channels with aspect ratio 0.6 < B_{cell}/d_{barrel} < 13

where V_x is the time-averaged longitudinal velocity component, V_{max} is the maximum velocity observed next to the free-surface, z is the vertical elevation measured above the invert, and d is the water depth.

For a two-dimensional flow,[10] the equation of conservation of mass gives a relationship between the bulk velocity V_{mean} and maximum free-surface velocity V_{max} since:

$$q = V_{mean} \times d = \int_{z=0}^{d} V_x \times dz = \frac{N}{N+1} \times V_{max} \times d \qquad (4.9)$$

with q the discharge per unit width, also called the unit discharge.

For a 1/N-th power law velocity distribution in a two-dimensional fully developed flow, a monotonic relationship may be derived in terms of the percentage of flow area, where the local time-averaged velocity V_x is less than U_{fish}:

$$A = \frac{LVZ}{B_{cell} \times d_{barrel}} = 100^{1-N} \times \left(\frac{N}{N+1}\right)^N \times \left(\frac{U_{fish}}{V_{mean}}\right)^N \qquad (4.10)$$

with A the relative LVZ area in the culvert barrel, defined as $A = LVZ/(B_{cell} \times d_{barrel})$ and expressed in percentage and (U_{fish}/V_{mean}) in percentage.[11] Equation (4.10) is shown for $N = 4.5$ in Figure 4.5 and compared to physical experimental data and 3D CFD numerical data in relatively narrow flumes $(0.6 < B_{cell}/d_{barrel} < 13)$.

In the following sections, Equation (4.10) for $N = 4.5$ is recommended to estimate the relative LVZ area because it compares favorably with detailed experimental and numerical CFD data (Fig. 4.5), and it is based upon a physically based principle (i.e. the equation of conservation of mass). The proposed design method shows implicitly a number of advantages. First, the relationship (Fig. 4.5, Equation (4.10)) is independent of hydrological implications, which could vary upon the requirements of different councils and sites. Second, the result is independent of the barrel culvert cell size. Third, the relationship is further independent of the downstream tailwater conditions.

4.5.5   Corner LVZ dimension

Since most small fish swim next to the bottom corners and along the sidewalls, the size of the LVZ in the bottom corners must be large enough to host at least one individual fish. For weak-swimming fish species, it is recommended to ensure LVZs with a minimum area of h_f by h_f (width by height) at the bottom corners of the culvert barrel cells, with h_f the vertical height (m) of the targeted fish species (Fig. 4.2). That is, the 45-degree diagonal from the corner must be larger than $2^{1/2} \times h_f$:

$$DL > 2^{1/2} \times h_f \tag{4.11}$$

with DL the 45-degree diagonal from the corner of the LVZ (Fig. 4.4). A larger value of the 45-degree diagonal from the corner might be specified for different targeted fish species. The selection of the bisector dimension derives directly from basic hydrodynamic considerations, since the axial vorticity component of the mean motion is zero on the bisector (Liggett *et al.*, 1965; Gessner, 1973).

Basic dimensional considerations show that the length DL of the 45-degree diagonal from the bottom corner is a function of the targeted velocity $V_x = U_{fish}$ relative to the bulk velocity V_{mean}, and to the aspect ratio B_{cell}/d_{barrel} of the barrel cell flow:

$$\frac{DL}{d} = F\left(\frac{U_{fish}}{V_{mean}}, \frac{B_{cell}}{d_{barrel}}\right) \tag{4.12}$$

Figure 4.6 presents a summary of detailed physical and 3D CFD numerical data, showing the dimensionless length DL/d_{barrel} as a function of the relative longitudinal velocity U_{fish}/V_{mean} and channel aspect ratio B_{cell}/d_{barrel}.

The analysis of these detailed physical data and CFD numerical data shows that the length DL of the 45-degree diagonal from the bottom corner may be correlated as:

$$\frac{DL}{d_{barrel}} = \sqrt{2} \times \exp\left(-\alpha \times \left(1 - \beta \times \frac{U_{fish}}{V_{mean}}\right)\right) \tag{4.13a}$$

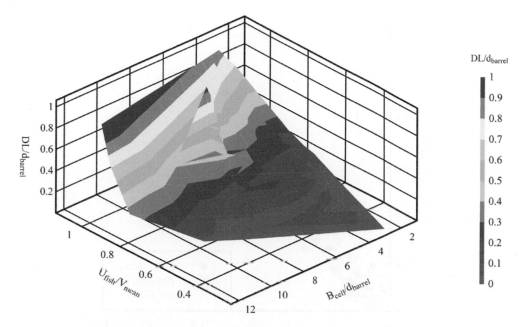

Figure 4.6 Relationship between dimensionless diagonal length DL/d_{barrel}, relative longitudinal velocity U_{fish}/V_{mean}, and aspect ratio B/d in smooth rectangular channels – surface plot of detailed physical experiments (Nezu and Rodi, 1985; Xie, 1998; Cabonce *et al.*, 2017, 2018) and CFD numerical results (Naot and Rodi, 1982; Leng and Chanson, 2018), with aspect ratio: 1 < B_{cell}/d_{barrel} < 12.5

with

$$\alpha = \frac{5.24}{1 - 0.8 \times \exp\left(-0.54 \times \dfrac{B_{cell}}{d_{barrel}}\right)} \tag{4.13b}$$

$$\beta = \left(1 + \frac{1}{5.82 + 0.000154 \times \left(\dfrac{B_{cell}}{d_{barrel}}\right)^{6.24}}\right)^{-1} \tag{4.13c}$$

A comparison between data and correlation is shown in Figure 4.7.

In summary, Equation (4.13) may be applied to predict the 45-degree diagonal length DL of the LVZ, and the diagonal length DL must fulfill Equation (4.11) for small-bodied fish.

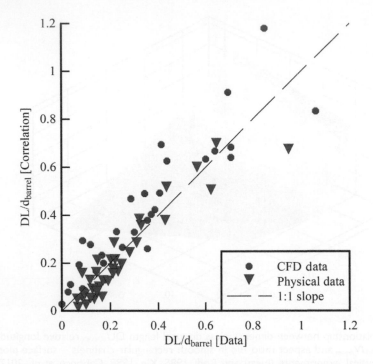

Figure 4.7 Relationship between estimated dimensionless diagonal length DL/d and calculated one in smooth rectangular channels (Eq. (4.13)) – data sets: detailed physical experiments (Nezu and Rodi, 1985; Xie, 1998; Cabonce *et al.*, 2017, 2019) and CFD numerical results (Naot and Rodi, 1982; Leng and Chanson, 2018), with aspect ratio $1 < B_{cell}/d_{barrel} < 12.5$

4.6   Discussion

Considering the design methodology introduced in the previous sections, the iterative procedure for the hydraulic design of a fish-friendly standard box culvert with smooth boundaries may be simplified, as shown in Figure 4.8.

Calculations are first conducted for design flow conditions. The optimum design for Q_{des} with a maximum acceptable afflux h_{max} would basically yield the minimum number of boxes $(N_{cell})_{des}$ required to achieve inlet control for a multicell culvert (Chapter 2). Two alternative methods may be used, being an engineering design nomograph (Concrete Pipe Association of Australasia, 2012) or a set of theoretical equations based on critical conditions and conservation of energy (Henderson, 1966; Chanson, 2004). Then the design must be checked against the outlet control situation, using the outlet control nomograph (Concrete Pipe Association of Australasia, 2012). The nomograph will give the afflux for the calculated number of cells at the outlet control. Ultimately, whichever afflux is bigger controls the flow, for example, if afflux for inlet control is larger than the afflux for outlet control, inlet control is achieved; otherwise, the number of cells must be increased, and outlet control calculations are repeated until the afflux is less than the maximum acceptable afflux h_{max}.

In the next stage, hydraulic calculations are performed for the culvert barrel at $Q < Q_T$. Since the most difficult flow conditions for upstream passage of small-body-mass fish occur

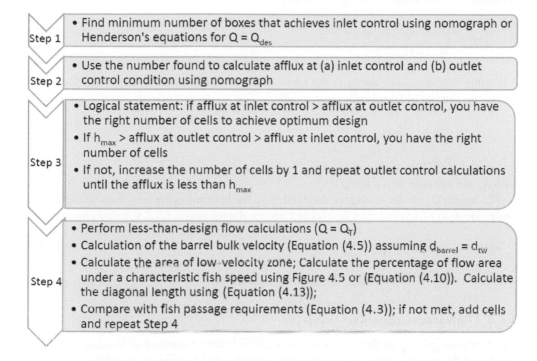

Step 1
- Find minimum number of boxes that achieves inlet control using nomograph or Henderson's equations for Q = Q$_{des}$

Step 2
- Use the number found to calculate afflux at (a) inlet control and (b) outlet control condition using nomograph

Step 3
- Logical statement: if afflux at inlet control > afflux at outlet control, you have the right number of cells to achieve optimum design
- If h$_{max}$ > afflux at outlet control > afflux at inlet control, you have the right number of cells
- If not, increase the number of cells by 1 and repeat outlet control calculations until the afflux is less than h$_{max}$

Step 4
- Perform less-than-design flow calculations (Q = Q$_T$)
- Calculation of the barrel bulk velocity (Equation (4.5)) assuming d$_{barrel}$ = d$_{tw}$
- Calculate the area of low-velocity zone; Calculate the percentage of flow area under a characteristic fish speed using Figure 4.5 or (Equation (4.10)). Calculate the diagonal length using (Equation (4.13));
- Compare with fish passage requirements (Equation (4.3)); if not met, add cells and repeat Step 4

Figure 4.8 Simplified flow chart of the optimum design at fish-friendly box culverts

for Q = Q$_T$, the calculations for upstream fish passage are typically focused on Q = Q$_T$. The bulk velocity in the culvert barrel is calculated using Equation (4.7). The flow area where V$_x$ < U$_{fish}$ may be estimated using Figure 4.5 or Equation (4.10). Typically, at least 15% of the flow area must experience longitudinal velocities under the critical fish swimming speed U$_{fish}$. Also, a minimum area of the LVZ at the barrel bottom corners must be fulfilled to pass the fish bodies, that is, DL > 2$^{1/2}$×h$_f$, with DL being calculated using Equation (4.13). If the results do not satisfy the required criteria of fish-friendly design, the number of cells must be revised by adding a further cell, and the modeling is to be repeated with an updated number of culvert barrel cells.

4.6.1   Commentary

The design calculations for upstream fish passage at Q = Q$_T$ may be undertaken in a slightly different manner with the same results.

To achieve LVZs corresponding to 15% of the flow, the ratio of characteristic fish swimming speed to bulk velocity must be less than 0.802, or 80.2% (Fig. 4.5 and Equation (4.10)). In other terms, the barrel bulk velocity V$_{mean}$ must satisfy:

$$V_{mean} < \frac{U_{fish}}{0.802} \tag{4.14}$$

By continuity, the water depth d_{barrel} in the culvert barrel for $Q = Q_T$ must fulfill:

$$d_{barrel} > \frac{Q_T}{N_{cell} \times B_{cell} \times \frac{U_{fish}}{0.802}} \tag{4.15}$$

Assuming that the water depth in the barrel is equal to the tailwater depth d_{tw} (Equation (4.5)), the number of barrel cells must satisfy:

$$N_{cell} > \frac{Q_T}{\frac{U_{fish}}{0.802} \times d_{tw} \times B_{cell}} \tag{4.16}$$

with d_{tw} the tailwater depth for $Q = Q_T$.

The length DL of the 45-degree diagonal from the bottom corner may then be calculated using Equation (4.13) and may be checked against the minimum LVZ size requirement:

$$DL > 2^{1/2} \times h_f \tag{4.11}$$

with h_f the vertical height (m) of the targeted fish species.

4.6.2 Important considerations

The proposed guidelines developed herein are based upon a number of basic assumptions (Section 4.5), and they apply to standard box culverts only. When one or more assumptions are untrue, the simplified guidelines should not be applied and complete detailed hydraulic engineering calculations should be conducted, following the design methodology developed in Sections 4.2 and 4.3. Similarly, if the culvert barrel shape is not a rectangular cross-section, the culvert should not be designed using the methodology and guidelines presented in Chapter 4.[12] Instead, a complete hydraulic modeling, using physical and/or numerical CFD, must be undertaken for the barrel configuration for both design and less-than-design discharges.

For standard box culverts, the design engineers must check that assumptions 1 and 2 are valid for the culvert structure. That is, first, the flood plain operates with subcritical flow and second, the free-surface elevation in the barrel equals the tailwater free-surface elevation. When assumption 2 is invalid, full hydrodynamic calculations must be conducted for less-than-design discharges.

Finally, it must be stressed that the design of a culvert intended to be constructed should require the certification of a professional civil engineer.

Notes

1 Inlet control means that the hydraulic control is located at the entrance (e.g. critical flow conditions take place in the barrel with free-surface inlet). Outlet control implies that the culvert flow is controlled at the outlet (i.e. by the tailwater conditions). See the glossary of technical terms in Appendix A.

2 Two different studies rarely use the same test methods and protocol, and the output is often a single-point measurement or a bulk velocity.

3 For example, small-bodied fish ($L_f < 100$ mm) and juveniles of larger fish.

4 This approach assumes implicitly that the discharge and velocity field are identical in all culvert barrel cells.

5 Leng *et al.* (2019) tested percentages of flow area corresponding to LVZs between 10% and 20%. The results suggested that 10% was too low, whereas there were relatively small differences between 15% and 20%. Thus, 15% of the flow area may be considered a robust target.

6 See the glossary of technical terms in Appendix A.

7 Also called the continuity principle.

8 Assuming implicitly that $Q_{cell} = Q/N_{cell}$, with N_{cell} the number of identical culvert barrel cells.

9 That is, the ratio of LVZ area to total wetted area is greater than 15%, in line with recent hydrodynamic data (Wang and Chanson 2018b, Cabonce *et al.* 2019), with the LVZ area being defined as the wetted area where the local time-averaged velocities are less than the characteristic fish swimming speed U_{fish}, with $0 < V_x < U_{fish}$.

10 That is, a wide open channel flow.

11 For example, if $U_{fish}/V_{mean} = 0.60 = 60\%$, use $(U_{fish}/V_{mean}) = 60$ in Equation (4.10). The result is $A = 4.07$, corresponding to $LVZ/(B_{cell} \times d_{barrel}) = 4.07\%$.

12 For completeness, the inlet control or outlet control nomographs presented in Chapter 2 apply only to standard box culverts.

Chapter 5

Design application

5.1 Presentation

A culvert design may vary from a simple geometry (standard culvert) to a hydraulically smooth shape (minimum energy loss [MEL] culvert). Considering the simple case of a standard box culvert (Fig. 5.1), the hydraulic structure consists of three components: the intake or inlet, the barrel or throat, and the diffuser or outlet. Based upon current design practices, the hydraulic characteristics of the culvert are the design discharge, the corresponding water depth in the natural stream in the absence of the culvert structure, and the maximum acceptable afflux (Chapter 2).

In terms of the upstream passage of fish in box culverts, several field observations (Behlke et al., 1991; Blank, 2008; Goettel et al., 2015) and near-full-scale experiments (Gardner, 2006; Wang et al., 2016a; Cabonce et al., 2017, 2019) reported fish seeking low-velocity zones (LVZs) associated with high-turbulence-intensity levels, typically next to the bottom corners and along the sidewalls, to pass through box culverts structures (see also movies in Appendix F). Very detailed velocity measurements in subcritical flows typical

(A) Single box culvert outlet along Le Rat river at Le Point Barré, near Pléboulle (22), France, on 28 June 2019

Figure 5.1 Standard box culvert design

(B) Multicell box culvert inlet beneath Caloundra Road, Caloundra QLD, Australia, on 10 October 2018

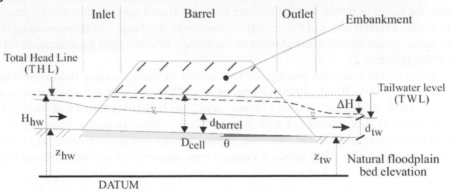

(C) Definition sketch of a box culvert

Figure 5.1 (Continued)

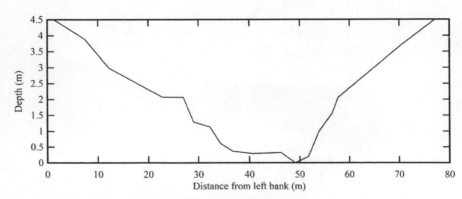

(A) Channel cross-section surveyed on 2 November 2010

Figure 5.2 River cross-section and hydrographic data of the Laura river at Laura NSW, Australia – Gauge Site 418021, Gwydir Catchment (data courtesy of WaterNSW)

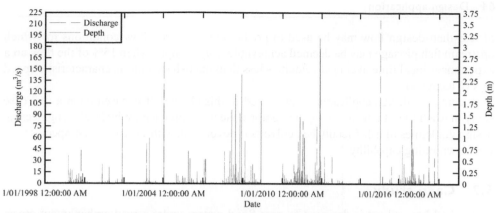

(B) Discharge data from 1 November 1998 to 1 November 2018 – catchment area: 311 km²

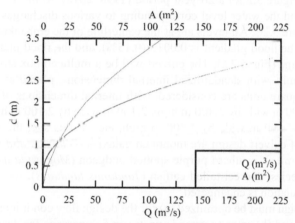

(C) Observations of (gauged) discharge Q and water depth between 1 November 1998 and 1 November 2018 (blue data) – relationship between cross-section area A and water depth (red curve)

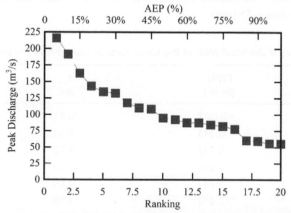

(D) Peak discharges for the period between 1 November 1998 and 1 November 2018 – flood frequency analysis based upon peak-over-threshold data series

Figure 5.2 (Continued)

of less-than-design flow may be used to predict culvert barrel flow conditions for which upstream fish passage may be deemed achievable: for example, when 15% of the flow area experiences local time-averaged velocities less than the fish swimming characteristic speed U_{fish} (Chapter 4).

A complete design application is presented in this chapter. It is based upon a real case study in Australia.[1] In the following paragraphs, the new approach for hydraulic engineering design guidelines of fish-friendly box culverts is developed, with a focus on fish species with weak swimming capability.

5.2   Case study

A standard box culvert is designed to pass flood waters under a road embankment crossing the Laura river flood plain at Laura NSW, Australia. The hydrological data of the site are regrouped in Figure 5.2 for a 20-year period (1998–2018). At the site, the ground level is −0.257 m RL and the water level corresponding to various discharges is summarized in Table 5.1 and Figure 5.2C. All levels are at the centerline of the embankment, which is 8 m wide at its base. The flood gradient is 0.0015 (0.15%), and the flood plain has an irregular cross-sectional shape (Fig. 5.2A). The culvert will be a multicell box structure, built using precast concrete units with standardized internal dimensions, placed at ground level. For simplicity, only square units are considered, with internal dimensions of 2.700 m wide by 2.700 m high, 2.400 m wide by 2.400 m high, 2.100 m wide by 2.100 m high, 1.800 m wide by 1.800 m high, 1.500 m wide by 1.500 m high, etc. At this site, the most relevant fish species in terms of culvert design are mountain galaxias (*Galaxias olidus*), river blackfish (*Gadopsis marmoratus*), southern purple spotted gudgeon (*Mogurnda adspersa*), which is an endangered species, and eel-tailed catfish (*Tandanus tandanus*), for which the Murray Darling basin population is endangered.

The culvert design must be optimized (a) for the design flow conditions with a maximum afflux of 0.450 m and (b) so as to provide upstream fish passage for discharges up to 10% of the design discharge (i.e. $Q_T/Q_{des} = 0.1$) to a guild of small-body-mass fish with a characteristic swimming speed U_{fish} of 0.36 m/s, corresponding to eel-tailed catfish (*Tandanus tandanus*) (Australia).[2] The bottom corner LVZ must be a minimum of 25 mm × 25 mm (i.e. $DL > 2^{1/2} \times 25$ mm = 35 mm).

Table 5.1 Rating curve of the flood plain of the Laura river at Laura NSW (Australia)

Q (m³/s)	TWRL (m RL)	d (m)	A (m²)
0.0063	−0.007	0.25	0.684
0.71	0.243	0.50	3.93
11.7	0.743	1.00	13.67
37.3	1.243	1.50	26.05
75.8	1.743	2.00	40.6
130.8	2.243	2.50	59.5
206	2.743	3.00	83.5

Notes: Tailwater rating level (TWRL) at the centerline of the embankment; A: flow cross-sectional area; Laura river gauge data at Laura NSW, Australia.

Develop the culvert design calculations for Q_{des} = 150 m³/s, 95 m³/s, 75 m³/s, and 55 m³/s. These design discharges correspond approximately to an annual exceedance probability (AEP) of 0.181, 0.283, 0.392, and 0.632, respectively.[3] The design flow conditions may be compared to the time series of gauged data (Fig. 5.2B) and the peak discharges for the same 20-year data set (Fig. 5.2D). The rating curve of the natural flood plain with its irregular cross-section area is given in Figure 5.2C and Table 5.1.

5.3   Detailed application

5.3.1   Design flow calculations

The calculations are first developed for design flow conditions. The barrel size is selected by a test-and-trial procedure in which both inlet control and outlet control calculations are performed for design flow conditions (Chapter 2). At the end, the optimum size is the smallest barrel size allowing for an afflux less than the maximum acceptable afflux (0.450 m) at design discharge (Herr and Bossy, 1965; Chanson, 2004). The full calculations are presented for Q_{des} = 150 m³/s, and the results for Q_{des} = 150 m³/s, 95 m³/s, 75 m³/s, and 55 m³/s are compared later.

5.3.1.1   Case Q_{des} = 150 m³/s

In the absence of a culvert and at design flow conditions, the flow depth in the flood plain is 2.64 m based upon the rating curve, and the corresponding specific energy is:

$$E = d + \frac{V^2}{2 \times g} = d + \frac{Q^2}{2 \times g \times A_{floodplain}^2} = 2.898 \text{ m} \tag{5.1}$$

with d the water depth, V the bulk velocity, and A the cross-section area of the flow.[4]

The flood plain flow is subcritical, since the water depth is greater than the critical flow depth, and the earlier flow conditions correspond to the tailwater flow conditions for both inlet and outlet control calculations: d_{tw} = 2.64 m and E_{tw} = 2.898 m.

With a 0.450 m afflux, the upstream or headwater depth is 0.450 m higher than the downstream tailwater depth d_{tw}:

$$d_{hw} = d_{tw} + \text{Afflux} = 3.09 \text{ m} \tag{5.2}$$

where the subscript hw refers to the headwater or upstream flow conditions.

First let us assume inlet control conditions. With an upstream flow depth d_{hw} = 3.09 m, the upstream-specific energy equals:

$$E_{hw} = H_{hw} - z_{hw} = d_{hw} + \frac{V_{hw}^2}{2 \times g} = d_{tw} + \frac{Q_{des}^2}{2 \times g \times A_{hw}^2} = 3.236 \text{ m} \tag{5.3}$$

Using a nomograph (Fig. 2.3), the inputs are:

Internal height	D_{cell} = 2.700 m
Upstream-specific energy	$H_{hw} - z_{hw}$ = 3.236 m
Headwater depth	$(H_{hw} - z_{hw})/D_{cell}$ = 1.20
45-degree wing walls	

The nomograph gives a discharge per unit width: $q_{cell} \approx 9.1$ m²/s (Fig. 5.3A). Thus, the barrel internal width is:

$$B_{min} = \frac{Q_{des}}{q_{cell}} = 16.5\,m \tag{5.4}$$

Since the barrel consists of identical precast concrete boxes, the number of cells is the smallest integer value N_{cell} fulfilling:

$$N_{cell} > \frac{B_{min}}{B_{cell}} \tag{5.5}$$

The result yields $N_{cell} = 7$. The internal barrel width is then $N_{cell} \times B_{cell} = 18.9$ m. The inlet control calculations can be recalculated for the corresponding discharge per unit width:

$$q_{cell} = \frac{Q_{des}}{N_{cell} \times B_{cell}} = 7.94\,m^2/s \tag{5.6}$$

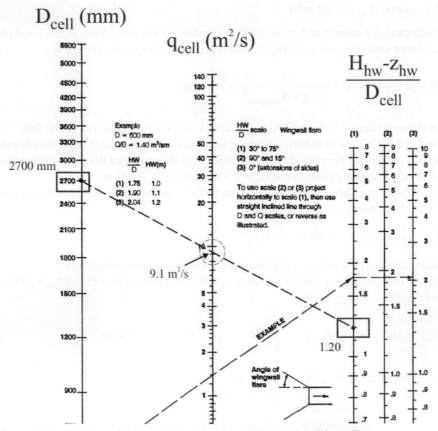

(A) Inlet control calculations for $Q_{des} = 150$ m³/s, $D_{cell} = 2.70$ m, and 0.450 m afflux

Figure 5.3 Design flow calculations – red frames are input data; blue frames are outputs

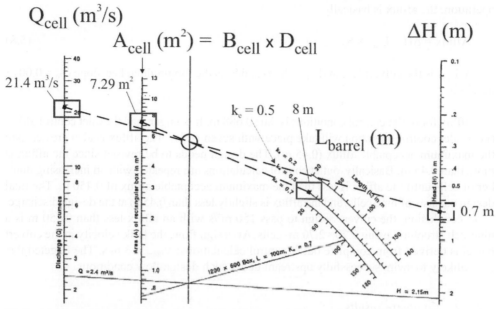

(B) Outlet control calculations for Q_{des} = 150 m³/s, D_{cell} = 2.70 m, and seven cells

Figure 5.3 (Continued)

The afflux at design discharge will be less than 0.45 m.[5]

In summary, for inlet control, the number of 2.700 m × 2.700 m precast cells is seven, since the design must be conservative. The free-board in the culvert barrel must be checked. At design discharge under inlet control conditions, the flow is critical in the barrel and the water depth is the critical flow depth:

$$d_c = \sqrt[3]{\frac{q_{cell}^2}{g}} = 1.86 \text{ m} \tag{5.7}$$

The free-board at design flow conditions is then as follows: Free-board = $D_{cell} - d_c$ = 2.70 – 1.86 = 0.84 m, corresponding to 31% clearance (i.e. free-board) between the water surface and roof (i.e. obvert).

Second, the outlet control calculations are performed for the design flow conditions with the seven-cell structure. Using the nomograph (Fig. 2.4), the inputs are:

Area of rectangular box	A_{cell} = 2.7 × 2.7 = 7.29 m²
Barrel length	L_{barrel} = 8 m
Entrance loss coefficient	k_e = 0.5
Discharge (per cell)	Q_{cell} = 150/7 = 21.4 m³/s

The outlet control nomograph gives a head loss $\Delta H = 0.7$ m (Fig. 5.3B). For outlet control operation, the afflux is basically:

$$\text{Afflux} = \Delta H - L_{culv} \times S_o \tag{5.8}$$

where L_{culv} is the culvert length ($L_{culv} = 8$ m) and S_o is the longitudinal bed slope ($S_o = 0.0015$ herein).

The afflux for outlet control conditions is thus 0.688 m. It is larger than the inlet control afflux (i.e. outlet control operation will take place with seven cells). The afflux is also greater than the maximum acceptable afflux (0.45 m). The design needs to be revised since the afflux is more than 0.45 m. Basically, outlet control calculations are repeated with an increasing number of cells until the afflux is less than the maximum acceptable afflux of 0.450 m. The final design consists of nine cells, and the afflux is slightly less than 0.45 m at the design discharge.

In conclusion, the culvert design to pass 150 m³/s with an afflux less than 0.450 m is a nine-cell structure, using 2.70×2.70 m² cells. At design flow, the bulk velocity in the culvert barrel is derived from complete outlet control calculations: $V_{mean} \approx 3$ m/s. The targeted fish are unlikely to swim successfully upstream under such design flow conditions.

5.3.1.2   Complete results

Complete hydraulic engineering design calculations are undertaken for $Q_{des} = 150$ m³/s, 95 m³/s, 75 m³/s, and 55 m³/s. The detailed results are summarized here:

	Units	Q_{des}				Comments
		150 m³/s	95 m³/s	75 m³/s	55 m³/s	
Q_{des}	m³/s	150.0	95.0	75.0	35.0	
AEP	–	0.181	0.283	0.392	0.632	of design discharge
d_{tw}	m	2.640	2.195	1.992	1.762	at design flow
$(N_{cell})_{des}$	–	9	7	6	6	
D_{cell}	m	2.700	2.400	2.400	2.100	Internal height
B_{cell}	m	2.700	2.400	2.400	2.100	Internal width
Flow regime	–	Outlet control	Outlet control	Outlet control	Inlet control	at design flow
Afflux	m	0.44	0.40	0.39	0.42	at design flow
$(N_{cell})_{des} \times A_{cell}$	m²	65.61	40.32	34.56	22.05	Total barrel cross-sectional area

The results show that the design discharge has a significant impact on the size of the culvert structure, namely the number of cells, size of the cells, and total barrel cross-sectional area. Note that the internal height of the barrel cell was selected to be larger than the tailwater depth to provide an adequate free-board during outlet control operation.

The final designs are outlined on the barrel channel cross-section in Figure 5.4. For $Q_{des} = 150$ m³/s, the culvert structure would be massive, and a bridge might be a more suitable design option.

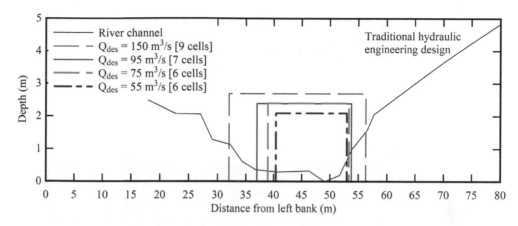

Figure 5.4 Dimensioned sketch of the standard box culvert design outlined for Q_{des} = 150 m³/s, 95 m³/s, 75 m³/s, and 55 m³/s (neglecting the precast unit wall thickness)

5.3.2 Culvert operation and fish passage for less-than-design flow

The fish swimming performance data may be linked to a minimum relative size of the LVZs (Chanson and Leng, 2018), where the local time-averaged longitudinal velocity component V_x satisfies:

$$0 < V_x < U_{fish} \tag{5.9}$$

with U_{fish} a characteristic fish speed, for example, set by a regulatory agency or based upon biological observations and swimming test data. In this section, hydrodynamic calculations are developed to predict the barrel flow conditions for which upstream fish passage may be deemed achievable, that is, when 15% of the flow area experiences local time-averaged velocities less than the characteristic fish swimming speed U_{fish} = 0.36 m/s.

Further, since most weak-swimming fish swim next to the bottom corners and along the sidewalls, the LVZs must have a minimum area of 25 mm by 25 mm (width by height) at the bottom corners of the culvert barrel cells, that is, 45-degree diagonal from corner (Fig. 5.5):

$$DL > 2^{1/2} \times 25 \text{ mm} \approx 35 \text{ mm} \tag{5.10}$$

The calculations of the barrel size for upstream fish passage in less-than-design flow conditions are iterative. They are typically conducted for the largest discharge for which upstream fish passage is required, since these flow conditions yield the fastest water velocities in the culvert barrel. That is, $Q = Q_T = 0.1 \times Q_{des}$ herein. The full calculations are developed for Q_{des} = 150 m³/s, and the results for Q_{des} = 150 m³/s, 95 m³/s, 75 m³/s, and 55 m³/s are discussed afterwards.

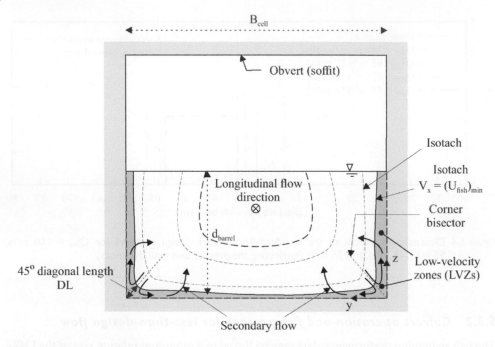

Figure 5.5 Definition sketch of LVZ and 45-degree diagonal from a corner in a box culvert barrel, looking downstream

5.3.2.1   Case Q_{des} = 150 m³/s

In the downstream flood plain, the flow conditions for $Q = Q_T$ are derived from the tailwater rating curve (Table 5.1, Fig. 5.2C). The tailwater flow conditions are:

Discharge	$Q = Q_T = 0.1 \times Q_{des}$ $= 15.0$ m³/s
Water depth	$d_{tw} = 1.085$ m
Cross-section area	$A_{tw} = 14.52$ m²
Bulk velocity	$V_{tw} = \dfrac{Q}{A_{tw}} = 1.033\,\text{m/s}$
Specific energy	$E_{tw} = d_{tw} + \dfrac{V_{tw}{}^2}{2 \times g} = 1.139\,\text{m}$

In the nine-cell culvert barrel, the discharge per unit width is 0.67 m²/s. Assuming that the barrel flow depth d_{barrel} is equal to the tailwater depth d_{tw}, the bulk velocity must satisfy the equation of conservation of mass:

$$V_{mean} = \frac{Q_{cell}}{B_{cell} \times d_{barrel}} = \frac{q_{cell}}{d_{barrel}} \tag{5.11}$$

where Q_{cell} is the water discharge in a single cell[6] and B_{cell} is the internal cell width. The bulk velocity in the barrel is $V_{mean} = 0.57$ m/s. Based upon detailed hydrodynamic data (Chapter 4, Fig. 4.5), the percentage of flow area where the local time-averaged velocities would be less than the characteristic swimming speed $U_{fish} = 0.36$ m/s would be about 5%.

Simply put, the culvert design with nine cells would be unsuitable to provide upstream fish passage to a guild of small-body-mass fish with a characteristic swimming speed of 0.36 m/s at 10% of the design discharge. The design needs to be revised with the inclusion of a larger number of cells. The calculations for $Q = Q_T$ (= $0.1 \times Q_{des}$ in this case study) are repeated with an increasing number of cells until the relative wetted area where the local time-averaged velocities is less than the characteristic swimming speed is 15% or more (Chapter 4).

After iterations, a culvert design with 12 cells operating at 10% of the design discharge gives a discharge per unit width of 0.463 m^2/s in the barrel, with a bulk velocity in the barrel $V_{mean} = 0.43$ m/s, and a relative LVZ area of 19%, where $V_x < U_{fish} = 0.36$ m/s. In the bottom corners of the barrel cells, the LVZs would have a 45-degree diagonal length DL from the corner of 0.24 m, well in excess of the minimum required 35 mm.

In conclusion, a revised standard box culvert design with 12 cells is capable of providing LVZs facilitating the upstream fish passage to a guild of small-bodied fish with a characteristic swimming speed U_{fish} of 0.36 m/s or more for discharges up to 15 m^3/s or $Q = Q_T = 0.1 \times Q_{des}$. This is achieved by ensuring that LVZ areas cover at least 15% of the water cross-sectional area. At design flow conditions $Q = Q_{des} = 150$ m^3/s, the 12-cell box culvert will operate under outlet control flow conditions and the afflux will be 0.24 m. That is, a reduction in afflux of 0.20 m at design discharge compared to the original nine-cell standard culvert design.

5.3.2.2 Complete results

Complete hydrodynamic calculations are performed for the three designs with $Q_{des} = 150$ m^3/s, 95 m^3/s, 75 m^3/s, and 35 m^3/s, with a characteristic swimming speed U_{fish} of 0.36 m/s. The detailed results are summarized here.

	Units	Q_{des}				Comments
		150 m³/s	*95 m³/s*	*75 m³/s*	*55 m³/s*	
Q_{des}	m³/s	150.0	95.0	75.0	55.0	
U_{fish}	m/s	0.36	0.36	0.36	0.36	Targeted fish species
N_{cell}	–	12	10	8	8	Fish-friendly design
D_{cell}	m	2.700	2.400	2.400	2.100	Internal height
B_{cell}	m	2.700	2.400	2.400	2.100	Internal width
$N_{cell} \times A_{cell}$	m²	87.5	57.6	46.1	35.3	Total barrel cross-sectional area
Afflux	m	0.24	0.20	0.20	0.42	at design flow: $Q = Q_{des}$
Q_T	m³/s	15	9.5	7.5	5.5	Threshold discharge
V_{mean}	m/s	0.43	0.42	0.44	0.42	in barrel for $Q = Q_T = 0.1 \times Q_{des}$
DL	m	0.24	0.23	0.17	0.20	Bisector dimension for $Q = Q_T = 0.1 \times Q_{des}$

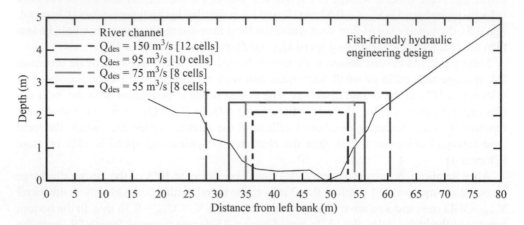

Figure 5.6 Dimensioned sketch of the fish-friendly standard box culvert barrel design outlined for Q_{des} = 150 m³/s, 95 m³/s, 75 m³/s, and 55 m³/s (neglecting the precast unit wall thickness)

The final designs of fish-friendly box culverts are outlined with the barrel cross-section in Figure 5.6. For Q_{des} = 150 m³/s and 95 m³/s, a bridge design would be a more suitable alternative to a massive multicell box culvert structure.

5.4   Commentary and discussion

These results were based upon the simplified design guidelines proposed in Chapter 4 (Section 4.5). They were compared to complete hydraulic engineering calculations. The comparison showed negligible quantitative differences in terms of culvert barrel depth and barrel bulk velocity for all three design discharges.[7]

Depending upon the characteristic swimming speed of the targeted fish species and guild, the current hydraulic engineering design guidelines of box culverts may or may not provide an adequate number of barrel cells to achieve upstream fish passage at 10% of the design discharge. Generally, the design calculations demonstrate conclusively that the cost of a fish-friendly box culvert increases with decreasing characteristic fish speed U_{fish}, increasing discharge threshold Q_T/Q_{des}, and increasing percentage of LVZ area (Chanson and Leng, 2018) (Chapter 4). All calculations showed the critical impact of the characteristic speed U_{fish} of targeted fish species. Culvert costs increase markedly in order to pass small-bodied fish at swimming speeds less than 0.3 m/s. For example, in the current design application, the selection of a characteristic swimming speed U_{fish} of 0.19 m/s, for southern purple spotted gudgeon (*Mogurnda adspersa*), an endangered fish species, was not considered, because it would lead to a huge, physically meaningless culvert structure, even for the smallest design discharge of 55 m³/s, and a bridge structure would be required[8] (Fig. 5.6). The effects of the characteristic swimming speed on the fish-friendly box culvert design and the size of the structure are illustrated in Figure 5.7. Figure 5.7 shows the dimensions of the case study structure, designed for three fish swimming speeds (U_{fish} = +∞, 0.36 m/s, and 0.19 m/s). Although the eight-cell fish-friendly box culvert would not allow upstream passage of southern purple spotted gudgeon up to the threshold discharge (herein $Q_T = 0.1 \times Q_{des}$), this does

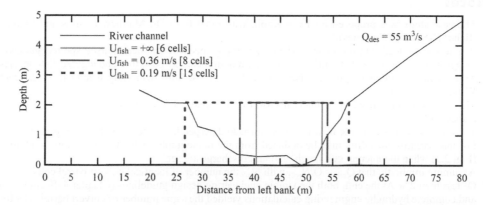

Figure 5.7 Dimensioned sketch of fish-friendly standard box culvert barrel designs outlined for Q_{des} = 55 m³/s for no consideration of fish passage (U_{fish} = +∞, U_{fish} = 0.36 m/s, and U_{fish} = 0.19 m/s)

not preclude the passage of the species during low-flow periods when the fish is more likely to be moving. Hydraulic calculations indicate that the eight-cell fish-friendly design would fulfil Equation (4.1) for southern purple spotted gudgeon up to Q/Q_{des} = 4.7%, corresponding to more than 91% of the catchment flow records for the past 20 years (Fig. 5.2).

By contrast, a targeted fish speed U_{fish} > 0.7 m/s seems much more achievable, but the biological implications are that most small-bodied fish would be blocked at these high water velocities.

The hydraulic engineering calculations presented a number of "unexpected" trends, rarely discussed in traditional culvert design guidelines. These are related to (a) the maximum acceptable afflux h_{max} and (b) tailwater rating curve. An increase in maximum acceptable afflux h_{max} yields an increase in upstream-specific energy, hence an increased bulk velocity in the barrel, at design discharge Q_{des} and a narrower barrel. In turn, the requirements for upstream fish passage are less likely to be met at less-than-design flow, that is, in particular at $Q = Q_T < Q_{des}$. Simply put, a substantial increase in the number of cells may be required with a large maximum acceptable afflux h_{max} for a given design discharge

The tailwater rating curve is the relationship between tailwater depth and discharge, or variations of natural downstream water level with water discharges (e.g. Fig. 5.2C). With outlet control operation at less-than-design discharges, a larger tailwater depth implies a slower fluid flow in the entire culvert barrel, and the fish passage requirement is more likely to be fulfilled. While these trends may be physically derived from basic hydrodynamic principles, they are rarely mentioned in current engineering design manuals of culverts because less-than-design water discharges are not specifically considered.

When the fish-friendly culvert design requires more cells than the optimum design for flood capacity (only), the revised design would operate with a smaller afflux at the design discharge. The reduction in upstream flooding might yield a lesser total cost of the structure (e.g. with a lower embankment and reduced impact on upstream catchment). The savings could contribute to offset partially the increased cost caused by the larger number of culvert barrel cells.

Moreover, the earlier example showed some impact of the design discharge on the final outcome. For Q_{des} = 150 m³/s, corresponding to an 18.1% AEP storm event, the final design would be a massive structure, and a bridge might be more appropriate than a culvert.

Notes

1 The case study was proposed by Evan Knoll, with input from Dr. Matthew Gordos and Marcus Riches (NSW DPI Fisheries).
2 It is acknowledged that the characteristic swimming speed of southern purple spotted gudgeon (*Mogurnda adspersa*) is 0.19 m/s. The design requirements to pass this endangered fish species would lead to a bridge design. See discussion in Section 5.4. Both swimming speed data come from Watson *et al.* (2019).
3 That is, an average recurrence interval (ARI) of 1:5 years, 1:3 years, 1:2 years, and 1:1 years, respectively.
4 The cross-section area of the flow is always measured perpendicular to the velocity direction.
5 For inlet control, the afflux may be deduced from the nomograph for inlet control flow conditions (Fig. 2.3) when the internal cell height and unit discharge are set.
6 Assuming implicitly that $Q_{cell} = Q/N_{cell}$, with N_{cell} the number of identical culvert barrel cells.
7 Of less than 2%. At the end, both the simple engineering design guidelines (Chapter 4, Section 4.5) and complete hydraulic engineering calculations yielded the same number of culvert barrel cells for the fish-friendly culvert designs.
8 Complete calculations led to a fish-friendly standard box culvert design with 15 cells to provide LVZs facilitating the upstream fish passage to small-bodied fish with a characteristic swimming speed U_{fish} of 0.19 m/s for discharges up to 5.5 m³/s or $Q = Q_T = 0.1 \times Q_{des}$.

Chapter 6

Discussion and practical considerations

6.1 General commentaries

A culvert structure typically incorporates an inlet, a barrel, and an outlet. Considering a box culvert with invert set at ground level and built in a flood plain with a mild slope, the culvert operation at less-than-design flows would include a relatively smooth flow convergence into the inlet leading to the barrel entrance. The application of the equations of conservation of mass and energy to the inlet flow implies that the water depth in the inlet decreases as the flow is accelerated, yielding maximum bulk velocity in the culvert barrel. At the downstream end of the barrel, the flow in the outlet experiences some deceleration, often associated with flow separation and large-scale turbulence, while the water depth would remain about the same as in the culvert barrel and downstream flood plain. A typical example of less-than design flow is discussed in Appendix B. Considering the upstream passage of small-bodied fish through a culvert operating at less-than-design discharges, the fish need to swim against a steady current and may encounter the largest fluid velocities in the culvert barrel at less-than-design flows. Conversely, the fish expand less mechanical work in the inlet and outlet, although the large-scale turbulent structures in the outlet might have some impact on fish swimming ability. This simple reasoning highlights that during less-than-design discharges, maximum velocities are experienced in the culvert barrel, thus justifying the present design approach focused on the box culvert barrel design.

In some cases, a lower invert may allow to retain a pool of water in the culvert barrel during dry to very low-flow conditions. A design for a culvert invert placed at ground level will result in conservative estimates compared to a recessed barrel invert, which are positives in terms of fish passage. Further, a lower barrel invert (or "wet" cell or low-flow channel) would require more advanced fluid dynamic calculations and likely is more expensive to build compared with a standard box culvert with barrel invert set at natural ground level (Leng *et al.*, 2019).

During the lifespan of a culvert, regular inspections and maintenance must be undertaken. Maintenance must be closely linked to inspection and design, as feedback from inspection and maintenance teams can reduce problems in future designs. Traditionally, works are conducted to prevent any reduction in discharge capacity at design flow conditions (e.g. obstructions are removed when they hinder the discharge capacity of the structure). The maintenance of culvert structures encompasses the removal of sediments and debris obstructing the waterway, in particular, upstream of the culvert and in the culvert inlet and barrel (Fig. 1.7). Known recurring issues include scour and erosion during flood events. Upstream fish passage requirements imply that a proper operation of the culvert at less-than-design discharge must be similarly considered during inspections and repair works. Culvert barrel roughening (e.g. concrete damage and algae growth) may increase locally the boundary roughness, hence the size of the low velocity zone (LVZ), with a potentially positive effect in terms of fish passage, when the roughening is regularly distributed. Sedimentation of culverts tends to affect adversely the culvert capacity during major floods, especially with cohesive sediments or when self-cleaning conditions are not achieved (QUDM, 2016). However, sediment infilling may also benefit some fish species (Duguay *et al.*, 2018).

In summary, the operation of box culverts with upstream fish passage capabilities during less-than-design discharges may imply a revised approach to maintenance and must be linked to the targeted fish species. The maintenance program has to be enlarged to ensure that the culvert operates adequately for less-than-design discharges with $Q < Q_T$, for example, that the LVZ and its longitudinal connectivity are not adversely affected by siltation and debris trapping.

6.2   Boundary treatments

A range of wall boundary treatments and appurtenances were recently tested (Fig. 6.1), with the aim being to improve the upstream passage of small fish in the culvert barrel, focusing on boundary conditions that have a minor impact on the discharge capacity of the culvert at

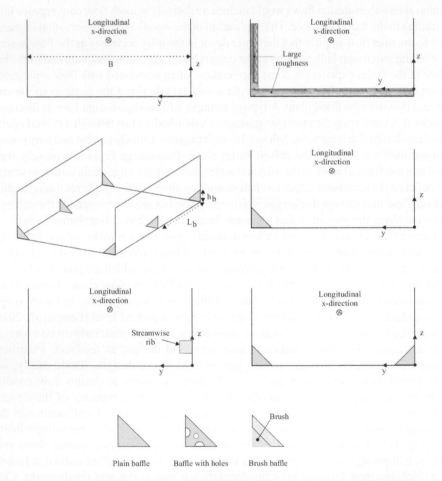

Figure 6.1 Examples of boundary treatments tested to improve the upstream passage of small-body-mass fish in a standard box culvert barrel (see Appendix E for detailed dimensions)

design discharge. Weak-swimming fish swim primarily next to the culvert barrel corners and sidewalls (Appendices C and F), although negative-wake flows might disorientate small-bodied fish (Cabonce *et al.*, 2018, 2019; Duguay *et al.*, 2018). LVZs most suitable to small-bodied fish passage must fulfil:

$$0 < V_x < U_{fish} \tag{6.1}$$

where V_x is the local time-averaged longitudinal velocity component and U_{fish} is a characteristic fish speed (Chanson and Leng, 2018). Fish navigability in a culvert barrel also depends on the connectivity between these low-positive-velocity zones (LPVZs), that is, where $0 < V_x < U_{fish}$.

The performances of various boundary treatments were compared in terms of the size of LPVZ: $0 < V_x < 0.5 \times V_{mean}$ and their longitudinal distribution and connectivity (Fig. 6.2). All the experimental works were conducted in rectangular channels with discharges typical of less-than-design discharges ($Q < Q_T$), and fish endurance tests were conducted for a limited range of configurations and discharges (Appendix E). Importantly, the comparison was developed with identical water discharge, in line with engineering design practices. Typical results are reported in Figure 6.2, showing the longitudinal distribution of the relative LPVZ area. In practical terms, most boundary treatments are associated with some form of limitations linked to installation, operation, and maintenance, ranging from drastic reduction in discharge capacity to siltation and failure. To date, most culvert operators favour smooth culverts without baffles, roughening, or appurtenances.

In the presence of different types of boundary treatments, all the observations showed the "sweet spots" that the fish exploit, namely regions of low positive velocity and high turbulence with intense secondary motion, irrespective of the boundary treatment. The comparative analysis of detailed hydrodynamic measurements with different boundary treatments

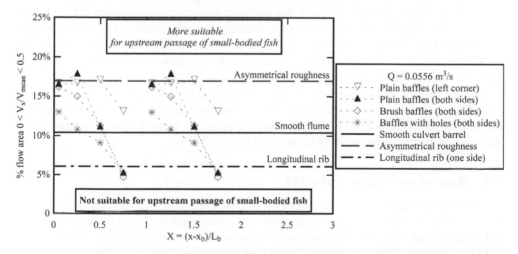

Figure 6.2 Longitudinal variation of fractions of LPVZ (i.e. $0 < V_x < 0.5 \times V_{mean}$) between different boundary treatments in a 12 m long and 0.5 m wide culvert barrel channel (Appendix E) – all data obtained for the same discharge $Q = 0.0556$ m³/s, over three longitudinal baffle spacings $(3 \times L_b)$

suggests that the requirements for a continuous, sizable LPVZ (e.g. $0 < V_x < 0.5 \times V_{mean}$) suitable to small-bodied fish might be best improved with an asymmetrically roughened culvert barrel,[1] with a roughness height comparable to the fish height (i.e. $k_s \sim h_f$).

6.2.1 Discussion

While leading scholars emphasized "the role of turbulence on biotic communities" (Maddock *et al.*, 2013, p. 433) and the complex "mechanics of fish–turbulence interactions" (Nikora *et al.*, 2003, p. 1380), what do we really know about turbulence? Professor Peter Bradsaw reminded us that "turbulence and its measurement are both controversial subjects" (Bradshaw, 1971, p. xii). Researchers cannot be complacent about turbulence, because "many of its seemingly simple questions remain unanswered" (Smits and Marusic, 2013, p. 25).

The interpretation of the turbulence typology is critical to any successful boundary treatment conducive to the upstream passage of small-bodied, weak-swimming fish (Chanson, 2019). A precise knowledge of the entire three-dimensional velocity field is essential, because the rate of work and energy required by a fish to thrust itself against the water discharge is proportional to the cube of the local fluid velocity, that is, V_x^3 (Wang and Chanson, 2018a). The in-depth understanding of the turbulent flow field constitutes a core requirement to comprehend the fish–fluid interactions and is a prerequisite for physically based mitigation measures of the ecological impact of culverts in terms of upstream fish passage.

6.3   Selection of the threshold discharge Q_T

The selection of the threshold discharge Q_T in relation to the design discharge Q_{des} has an important impact on the final design of a fish-friendly box culvert structure. In Chapter 5, the threshold discharge was selected at 10% of the design flow. For that catchment, the stream flow was less than the threshold discharge for between 95% and 98% of the 10-year hydrological data record, depending upon the design discharge selection. A limited study of eastern Australian catchments showed that the river discharge was less than $0.1 \times Q_{des}$ for more than 90% to 95% of the time for design discharges corresponding to an annual exceedance probability (AEP) of 0.392 to 0.632. The survey encompassed coastal and inland catchments, and the result was not found to be particularly sensitive to the catchment location, size, and type of hydrological event (Chanson and Leng, 2018).

In other regions, the climatic environment, hydrological conditions, and aquatic fauna could lead to the selection of a different relative threshold discharge Q_T/Q_{des}, for example, as discussed by Kilgore *et al.* (2010) for North America. In western Europe, a target is the allowance of fish passage for 300 days per year (DWA, 2014).

6.4   Box versus pipe culverts

Finally the present guidelines are developed for box culverts. While box culverts might be more effective in terms of fish passage, a large number of pipe culvert structures are installed worldwide, and the reader may ask: What are the key differences between box and pipe culvert barrels?

In box culverts, LVZs are closely linked to the secondary currents in the barrel corners and to the associated transverse gradient in local boundary shear stress. Circular pipe culverts operating at less-than-design flows present a vastly different velocity field compared to box culverts because of the smooth wetted perimeter geometry and the absence of corners. LVZs are thin smaller regions due to the lack of transverse boundary shear

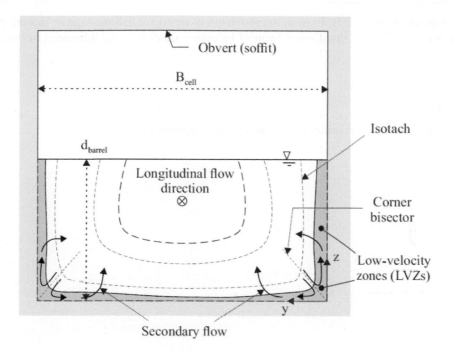

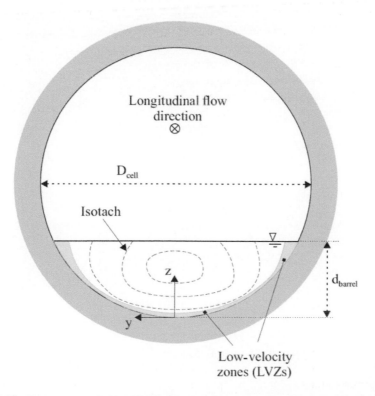

Figure 6.3 Comparative sketch of LVZ in box and pipe culvert flow fields for less-than-design discharges

stress gradients (Fig. 6.3). In turn, the provision of upstream fish passage in smooth pipe culverts is unlikely without some form of boundary treatment and advanced turbulence manipulation.

Note

1 That is, better than with baffles or streamwise rib(s) (Appendix E).

Chapter 7

Conclusion

A box culvert is a covered rectangular channel designed to pass water through an embankment (Fig. 7.1). The recognition of the adverse ecological impacts of culverts on upstream fish passage is driving the development of new standard box culvert design guidelines, with a focus on weak-swimming fish species seeking LVZs to minimize energy expenditure. In the context of Australian fish passage at road crossings and culverts, it is recognized that fish can swim through culverts as soon as the water discharges in the culvert barrel (i.e. $Q > 0$) and the water depth is suitable to the targeted fish species. Most fish swim upstream preferentially in the corners and near the sidewalls of the box culvert barrel, as observed with a number of Australian and North American fish species.

New hydraulic engineering guidelines are proposed based upon a number of basic design considerations that are relevant to most standard box culverts and fish species:

1 Design optimization for flood capacity for $Q = Q_{des}$ and for upstream fish passage for $Q < Q_T$
2 Provision of an LVZ representing at least 15% of the flow area and where $0 < V_x < U_{fish}$ for upstream fish passage (i.e. $Q < Q_T$), where U_{fish} is a characteristic fish speed, for example, set by a regulatory agency based upon biological observations and swimming test data
3 A smooth standard box culvert design, without appurtenance

The novel approach relies upon a solid understanding of turbulence in the box culvert barrel at less-than-design discharges and an accurate, physically based knowledge of the entire velocity field in the culvert barrel, specifically the longitudinal velocity map, to accurately characterize the LVZ next to the barrel walls and corners. Although the focus of the present guidelines is on the upstream passage of weak-swimming fish, including small-body-mass fish and juveniles of larger fish, the approach and methodology are relevant to most standard box culvert structures and can be applied to a very broad range of fish species.

The fundamental design considerations lead to a two-stage hydraulic engineering design. First, the minimum number of culvert barrel cells[1] is calculated to achieve inlet control at design flow conditions, based upon current standards for optimum flood capacity design at the culvert site. Second, considerations for upstream fish passage are embedded into the design method, using biological considerations. The revised fish-friendly culvert design may include a larger number of barrel cells than the original design. In such a case, the reduction in upstream flooding might contribute to some savings, which could partially offset the increased cost caused by the larger culvert barrel dimensions. Finally, the operation of box

(A) Masonry box culvert outlet at Erquy, France, on 28 June 2019 – 19th-century construction

(B) Standard box culvert inlet and barrel entrance along West creek beneath Herries Street, Toowoomba QLD, Australia, on 10 March 2011

Figure 7.1 Standard box culvert structures

culverts with upstream fish passage capabilities during less-than-design flows must imply a revised approach to maintenance, linked to the targeted fish species.

The present monograph develops a physically based rationale for a fish-friendly standard box culvert design, embedding state-of-the-art hydrodynamic calculations into current hydraulic engineering design methods to yield cost-effective outcomes. By bridging the gap between engineering and biology, this novel approach may contribute to the restoration of catchment connectivity for small-bodied native fish. Finally, it must be stressed that the design of a culvert that is intended to be constructed would require the certification of a professional civil engineer.

Note

1 Or the smallest internal barrel dimensions for a single-cell standard box culvert.

Appendices

Glossary of technical terms

Abutment   part of the valley side against which the dam or bridge is constructed.

Accretion   increase of channel bed elevation resulting from the accumulation of sediment deposits.

Advection   movement of a mass of fluid that causes a change in temperature or in other physical or chemical properties of a fluid.

AEP   probability of exceedance of a given discharge within a period of one year, generally expressed as a percentage. The AEP is estimated by extracting the annual maximum in each year to produce an annual maxima series (AMS) (Ball *et al.*, 2016).

Aerobic   activity involving   indeed, needing – oxygen Aerobic fish swimming can be maintained for an extended period without fatigue, and metabolic activity utilizes only red muscle tissues.

Afflux   rise of water level above normal level (i.e. natural flood level) on the upstream side of a culvert or of an obstruction in a channel. In the United States, it is commonly referred to as maximum backwater.

Aggradation   rise in channel bed elevation caused by deposition of sediment material. Another term is accretion.

Aleatory uncertainty   also called inherent uncertainty, it refers to the uncertainty that arises through natural randomness or natural variability observed in nature (Ball *et al.*, 2016).

Allowable soil pressure   maximum pressure permitted on the foundation soil with appropriate safety against rupture of the soil mass or movement leading to the structure impairment. It is also called allowable bearing.

Alternate depth   In open channel flow, for a given flow rate and channel geometry, the relationship between the specific energy and flow depth indicates that for a given specific energy, there is no real solution (i.e. no possible flow), one solution (i.e. critical flow), or two solutions for the flow depth. In the latter case, the two flow depths are called alternate depths: one corresponds to a subcritical flow and the second to a supercritical flow.

Amphidromous fish   diadromous fish migrating between freshwater and the sea at some stages of the life cycle other than breeding (e.g. forked-tail catfish).

Anadromous fish   diadromous fish spending most of their lives at sea and migrating to freshwater to breed (e.g. lamprey).

Anaerobic   activity not requiring oxygen. Anaerobic fish swimming cannot be maintained for an extended period, and metabolic activity utilizes only white muscle tissue.

Analytical model   system of mathematical equations that are the algebraic solutions of the fundamental equations.

Anguiliform fish propulsion   pure undulatory mode of fish propulsion in which the whole fish body contributes, although the amplitude of undulation may increase toward the tail (Lighthill, 1975). Named after the common eel (*Anguilla*).

Apelt   Colin J. Apelt is an emeritus professor in civil engineering at the University of Queensland (Australia).

Apron   the area at the downstream end of a weir or culvert outlet to protect against erosion and scouring by water.

ARI   average value of the period between exceedances of a given discharge, expressed typically as 1 in Y years. The ARI is derived from a peak-over-threshold series (PoTS), where every value over a chosen threshold is extracted from the period of record (Ball *et al.*, 2016).

Armoring   progressive coarsening of the bed material resulting from the erosion of fine particles. The remaining coarse material layer forms an armor, preventing further bed erosion.

Backwater   In a subcritical flow, the longitudinal flow profile is controlled by the downstream flow conditions (e.g. an obstacle, a structure, a change of cross-section). Any downstream control structure (e.g. bridge piers, weir) induces a backwater effect. More generally the term backwater calculations or backwater profile refers to the calculation of the longitudinal free-surface profile in an open channel. The term is used for both supercritical and subcritical free-surface flows.

Backwater calculation   calculation of the free-surface profile in open channels. The first successful calculations were developed by the Frenchman J. B. Bélanger, who used a finite difference step method for integrating the equations (Belanger, 1828).

Barré de Saint-Venant   Adhémar Jean Claude Barré de Saint-Venant (1797–1886), French engineer of the Corps des Ponts et Chaussées, who developed the equation of motion of a fluid particle in terms of the shear and normal forces exerted on it (Barré de Saint-Venant, 1871a, 1871b).

Barrel   for a culvert, the central section where the cross-section is at a minimum. Another term is the throat.

Baseflow   stream discharge hydrograph contributed primarily from the groundwater discharge.

Bed load   sediment material transported by rolling, sliding, and saltation motions along the bed.

Bélanger   Jean-Baptiste Ch. Bélanger (1789–1874) was a French hydraulician and professor at the Ecole Nationale Supérieure des Ponts et Chaussées (Paris). He first suggested the application of the momentum principle to hydraulic jump flow (Belanger, 1841).

Bélanger equation   momentum equation applied across a hydraulic jump in a horizontal rectangular channel and named after J.B.C. Bélanger (Chanson, 2009a).

Bernoulli   Daniel Bernoulli (1700–1782) was a Swiss mathematician, physicist, and botanist who developed the Bernoulli equation in his *Hydrodynamica, de viribus et motibus fluidorum* textbook (first draft in 1733, first publication in 1738, Strasbourg).

Bernoulli equation   basic principle derived from the Navier-Stokes equation, assuming no energy loss.

Borda   Jean-Charles de Borda (1733–1799) was a French mathematician and military engineer who investigated the flow through orifices and developed the Borda mouthpiece.

Bottom outlet   opening near the bottom of a dam for draining the reservoir and flushing out reservoir sediments.

Boundary layer   flow region next to a solid boundary where the flow field is affected by the presence of the boundary and where friction plays an essential part. A boundary layer flow is characterized by a range of velocities across the boundary layer region, from zero at the boundary to the free-stream velocity at the outer edge of the boundary layer.

Boussinesq   Joseph Valentin Boussinesq (1842–1929) was a French hydrodynamicist and professor at the Sorbonne University (Paris). His treatise "Essai sur la théorie des eaux courantes" (Boussinesq, 1897) is an outstanding contribution to the hydraulics literature.

Boussinesq coefficient   momentum correction coefficient named after J.V. Boussinesq, who first proposed it (Boussinesq, 1897).

Boussinesq-Favre wave   an undular surge (see *Undular surge*).

Boys   P.F.D. du Boys (1847–1924) was a French hydraulic engineer. He made a major contribution to the understanding of sediment transport and bed load transport (Boys, 1879).

Bresse   Jacques Antoine Charles Bresse (1822–1883) was a French applied mathematician and hydraulician. He was a professor at the Ecole Nationale Supérieure des Ponts et Chaussées (Paris) as a successor of J.B.C. Bélanger. His contribution to gradually varied flows in open channel hydraulics is considerable (Bresse, 1860).

Broad-crested weir   a weir with a flat long crest is called a broad-crested weir when the ratio of crest length to upstream head is greater than 1.5:3. When the crest is long enough, the pressure distribution along the crest is hydrostatic and the flow depth equals the critical flow depth $d_c = (q^2/g)^{1/3}$.

Buat   Comte Pierre Louis George du Buat (1734–1809) was a French military engineer and hydraulician and friend of Abbé C. Bossut. Du Buat is considered the pioneer of experimental hydraulics. His textbook (Buat, 1779) was a major contribution to flow resistance in pipes, open channel hydraulics, and sediment transport.

Bulk velocity   cross-sectional averaged velocity or mean flow velocity.

Byewash   channel to carry spilled or wasted waters (i.e. ancient name for a spillway).

Carangiform fish propulsion   fish propulsion mode in which the undulations are confined to a small posterior fraction of body length, mostly in the posterior third part of the fish's length (Lighthill, 1969; Blake, 1983). Named after *Caranx*, a genus of tropical to subtropical marine fishes in the jack family Carangidae.

Cartesian coordinate   one of three coordinates that locate a point in space and measure its distance from one of three intersecting coordinate planes measured parallel to that one of three straight-line axes that is the intersection of the other two planes. It is named after the French mathematician René Descartes.

Catadromous fish   diadromous fish spending most of their lives in freshwater and migrating to the sea to breed (e.g. barramundi).

Catchment   drainage basin.

Chézy   Antoine Chézy (1717–1798) (or Antoine de Chézy) was a French engineer and member of the French Corps des Ponts et Chaussées. He designed canals for the water supply of the city of Paris. In 1768, he proposed a resistance formula for open channel flows called the Chézy equation.

Chézy coefficient   resistance coefficient for open channel flows first introduced by the Frenchman Antoine Chézy. Although thought to be a constant, the coefficient is a function of the relative roughness and Reynolds number.

Choke   in open channel flow, a channel contraction might obstruct the flow and induce the appearance of critical flow conditions (i.e. control section). Such a constriction is sometimes called a "choke."

Choking flow   critical flow in a channel contraction. The term is used for both open channel flow and compressible flow.

Cofferdam   temporary structure enclosing all or part of the construction area so that construction can proceed in dry conditions. A diversion cofferdam diverts a stream into a pipe or channel.

Cohesive sediment   sediment material of very small sizes (i.e. less than 50 μm) for which cohesive bonds between particles (e.g. intermolecular forces) are significant and affect the material properties.

Colebrook-White formula   formula to calculate friction loss coefficients in pipes, tubes, and ducts.

Collars   see *Cutoff collars*.

Conjugate depth   in open channel flow, another name for sequent depth.

Consolidation   gradual reduction in the volume of a soil mass as a result of compressive stresses.

Control   considering an open channel, subcritical flows are controlled by the downstream conditions. This is called a downstream flow control. Conversely, supercritical flows are controlled only by the upstream flow conditions (i.e. upstream flow control).

Control section   in an open channel, the cross-section where critical flow conditions take place. The concepts of control, hydraulic control, and control section are used with the same meaning.

Control surface   the boundary of a control volume.

Control volume   refers to a region in space and is used in the analysis of situations where flow occurs into and out of the space.

Coriolis   Gustave Gaspard Coriolis (1792–1843) was a French mathematician and engineer of the Corps des Ponts et Chaussées who first described the Coriolis force (i.e. effect of motion on a rotating body).

Coriolis coefficient   kinetic energy correction coefficient named after G.G. Coriolis, who first introduced this velocity correction coefficient (Coriolis, 1836).

Critical depth   the flow depth for which the mean specific energy is at a minimum.

Critical flow conditions   in open channel flows, the flow conditions, such as the specific energy (of the mean flow), that are at a minimum. If the flow is critical, small changes in specific energy cause large changes in flow depth (Bakhmeteff, 1912; Chanson, 2006, 2008). In practice, critical flow over a long reach of channel is unstable.

Critical slope   when the uniform equilibrium flow depth is equal to the critical flow depth, the uniform equilibrium flow is critical, and the slope is called a critical slope. Critical slopes are seldom found in nature, because critical flow motion is unstable.

Culvert   covered channel of a relatively short length installed to pass water through an embankment (e.g. highway, railroad, dam).

Cutoff collars   cutoff collars are plates around culverts and conduits designed to minimize seepage and reduce the risks of piping in the embankment along the pipe outer shell (USBR, 1987).

Darcy   Henri Philibert Gaspard Darcy (1805–1858) was a French civil engineer. He studied at Ecole Polytechnique between 1821 and 1823, and later at the Ecole Nationale Supérieure des Ponts et Chaussées (Brown, 2002). He performed numerous experiments of flow resistance in pipes (Darcy, 1858) and in open channels (Darcy and Bazin, 1865) and of seepage flow in porous media (Darcy, 1856). He gave his name to the Darcy-Weisbach friction factor and to the Darcy law in porous media.

Darcy-Weisbach friction factor   dimensionless parameter characterizing the friction loss in a flow. It is named after the Frenchman H.P.G. Darcy and the German J. Weisbach.

Debris   comprises mainly large boulders, rock fragments, gravel-sized to clay-sized material, and tree and woody material that accumulate in creeks.

Degradation   lowering in channel bed elevation resulting from the erosion of sediments.

Diadromous fish   migratory fish species that migrate between freshwater and seawater at fixed seasons or life stages.

Dimensional analysis   organization technique used to reduce the complexity of a study by expressing the relevant parameters in terms of numerical magnitude and associated units and then grouping them into dimensionless numbers. The use of dimensionless numbers increases the generality of the results.

Diversion channel   waterway used to divert water from its natural course.

Drainage layer   layer of pervious material to relieve pore pressures and/or to facilitate drainage (e.g. drainage layer in an earthfill dam).

Drop structure   single-step structure characterized by a sudden decrease in bed elevation.

Dupuit   Arsène Jules Etienne Juvénal Dupuit (1804–1866) was a French engineer and economist.

Earth dam   massive earthen embankment with sloping faces and made watertight.

Eddy viscosity   a concept proposed by Boussinesq (1897) that characterizes the transport and dissipation of energy in the smaller-scale flow. It is another name for the momentum exchange coefficient, and it is also called the "eddy coefficient" by Schlichting (1979) (see *Momentum exchange coefficient*).

Effective rainfall   proportion of catchment rainfall that finds its way into a stream.

Embankment   fill material (e.g. earth, rock) placed with sloping sides and with a length greater than its height.

Epistemic uncertainty   refers to uncertainty associated with the state of knowledge of a physical system, the ability to measure it, and the inaccuracies in the predictions of the physical system (Ball *et al.*, 2016).

Explicit method   calculation containing only independent variables; numerical method in which the flow properties at one point are computed as functions of known flow conditions only.

Face   external surface that limits a structure: for example, water face (i.e. upstream face) of a weir.

Fawer jump   undular hydraulic jump.

Filter   layer(s) of pervious soil materials placed to provide drainage without movement of soil particles.

Finite differences   approximate solutions of partial differential equations that consist essentially of replacing each partial derivative by a ratio of differences between two immediate values: for example, $\partial V / \partial t \approx \delta V / \delta t$. The method was first introduced by Runge (1908).

First-order (or second-order) upwind scheme   a class of numerical discretization methods that discretize hyperbolic partial differential equations by using differencing biased in the direction determined by the sign of the characteristic speeds. First- or second-order denotes the order of numerical uncertainty.

Fixed-bed channel   the bed and sidewalls are nonerodible boundaries. Neither erosion nor accretion occurs.

Flash flood   flood of short duration with a relatively high peak flow rate.

Flood frequency   frequency with which a flood has the probability of recurring. Measures of the rarity of a rainfall event include the average recurrence interval (ARI) and annual exceedance probability (AEP). When ARI is expressed in years, the relationship between AEP and ARI is:

$$AEP = 1 - e^{-1/ARI}$$

Generally, it is preferable to express the rarity of an event in terms of AEP (Ball *et al.*, 2016).

Free-board   free-space clearance between the mean free-surface level and the roof (i.e. obvert). In a culvert, the free-board in the barrel must be at least 20% to prevent an adverse effect (Chanson, 2004, p. 445).

Free-surface   interface between a liquid and a gas. More generally a free-surface is the interface between the fluid (at rest or in motion) and the atmosphere. In two-phase gas–liquid flow, the free-surface region includes also the air–water interface of gas bubbles and liquid drops.

Free-surface aeration   natural aeration occurring at the free-surface of high-velocity flows (also referred to as self-aeration).

Froude   William Froude (1810–1879) was a English naval architect and hydrodynamicist who invented the dynamometer and used it for the testing of model ships in towing tanks. He was assisted by his son, Robert Edmund Froude, who, after the death of his father, continued some of his work. In 1868, he used Reech's law of similarity to study the resistance of model ships.

Froude number   proportional to the square root of the ratio of the inertial forces over the weight of fluid. The Froude number is used generally for scaling free-surface flows, open channels, and hydraulic structures. Although the dimensionless number was named after William Froude, several French researchers used it before (Dupuit, 1848; Bresse, 1860; Bazin, 1865). Ferdinand Reech introduced the dimensionless number for testing ships and propellers in 1852. The number is also called the Reech-Froude number.

Gate   valve or system for controlling the passage of a fluid. In open channels the two most common types of gates are the underflow gate and the overflow gate.

Gauckler   Philippe Gaspard Gauckler (1826–1905) was a French engineer and member of the French Corps des Ponts et Chaussées. He reanalyzed the experimental data for open channel flows of Darcy and Bazin (1865) and presented in 1867 a flow resistance formula (i.e. the Gauckler-Manning formula), too often improperly called the Manning equation (Gauckler, 1867).

G.K. formula   empirical resistance formula developed by the Swiss engineers E. Ganguillet and W.R. Kutter in 1869.

Gradually varied flow   characterized by relatively small changes in velocity and pressure distributions over a short distance (e.g. long waterway).

Headwater   upstream flow.

Headwater depth   upstream flow depth.

Headwater level   upstream free-surface elevation.

Hydraulic diameter   defined as the equivalent pipe diameter that is, four times the cross-section area divided by the wetted perimeter. The concept was first expressed by the Frenchman P.L.G. du Buat (Buat, 1779).

Hydraulic jump   transition from a rapid supercritical flow to a slow flow motion (i.e. subcritical flow). Although the hydraulic jump was described by Leonardo da Vinci, the first experimental investigations were published by Giorgio Bidone in 1820. The present theory of the jump was developed by Bélanger (1841) and verified experimentally since (e.g. Bakhmeteff and Matzke, 1936).

Hydrostatic pressure   pressure that is exerted by a fluid at equilibrium at a given position within the fluid caused by gravity.

Ideal fluid   frictionless and incompressible fluid. An ideal fluid has zero viscosity: that is, it cannot sustain shear stress at any point.

Implicit method   calculation in which the dependent variable and one or more independent variables are not separated on opposite sides of the equation; numerical method in which the flow properties at one point are computed as functions of both independent and dependent flow conditions.

Implicit solution   a numerical solution in the time domain that is solved by equations involving both the current state of the system and the later one.

Inflow   (1) upstream flow; (2) incoming flow.

Inlet   (1) upstream opening of a culvert, pipe, or channel; (2) a tidal inlet is a narrow water passage between peninsulas or islands.

Inlet control   in a culvert, inlet control flow conditions mean that the hydraulic control is located at the entrance: for example, critical flow conditions take place in the barrel with free-surface inlet.

Intake   any structure in a reservoir through which water can be drawn into a waterway or pipe. By extension, upstream end of a channel.

Internal friction   portion of a soil's shearing strength due to the interlocking of the soil grains and the resistance to motion between grains. It is also called shear resistance.

International system of units   see *Système international d'unités*.

Invert   (1) lowest portion of the internal cross-section of a conduit; (2) channel bed of a spillway; (3) bottom of a culvert barrel.

Inviscid flow   a nonviscous flow.

Ippen   Arthur Thomas Ippen (1907–1974) was a professor in hydrodynamics and hydraulic engineering at MIT (USA). Born in London of German parents and educated in Germany (Technische Hochschule in Aachen), he moved to the United States in 1932, where he obtained MS and PhD degrees at the California Institute of Technology. There he worked on high-speed free-surface flows with Theodore von Karman. In 1945 he was appointed at MIT until his retirement in 1973.

Irrotational flow   defined as a zero-vorticity flow. Fluid particles within a region have no rotation. If a frictionless fluid has no rotation at rest, any later motion of the fluid will be irrotational. In irrotational flow, each element of the moving fluid undergoes no net rotation, with respect to chosen coordinate axes, from one instant to another.

JHRC   jump height rating curve.

JHRL   jump height rating level.

Karman   Theodore von Karman (or von Kármán) (1881–1963) was a Hungarian fluid dynamicist and aerodynamicist who worked in Germany (1906–1929) and later in the United States. He was a student of Ludwig Prandtl in Germany. He gave his name to the vortex shedding behind a cylinder (i.e. Karman vortex street).

Karman constant (or von Karman constant)   pseudo-universal constant K of proportionality between the Prandtl mixing length and the distance from the boundary. Experimental results indicate that $K = 0.4$.

Kennedy   Professor John Fisher Kennedy (1933–1991) was a hydraulic professor at the University of Iowa, United States. He succeeded Hunter Rouse as head of the Iowa Institute of Hydraulic Research.

Keulegan   Garbis Hovannes Keulegan (1890–1989) was an Armenian mathematician who worked as hydraulician for the US Bureau of Standards since its creation in 1932.

Left abutment   abutment on the left-hand side of an observer when looking downstream.

Left bank (left wall)   looking downstream, the left bank or the left channel wall is on the left.

Lining   coating on a channel bed to provide water tightness, to prevent erosion, or to reduce friction.

Log law (or law of wall)   describes the boundary layer characteristics, that is, velocity as a function of distance from the wall, near the close vicinity of walls based upon a Prandtl mixing-length model (Schlichting, 1979; Chanson, 2014).

Low-velocity zone (LVZ)   flow area where the time-averaged longitudinal velocity V_x is small, typically substantially smaller than the bulk velocity V_{mean}. Low-velocity zones are essential for successful upstream passage because the rate of work and energy required by fish to thrust themselves against the current is proportional to the cube of the local fluid velocity (Wang and Chanson, 2018a). Recent work on small-body fish showed further that LVZs should not exhibit strong recirculation or negative velocity (Cabonce et al., 2018).

LVZ   see Low-velocity zone.

Manning   Robert Manning (1816–1897) was chief engineer of the Office of Public Works, Ireland. In 1889, he presented two formulas (Manning, 1890). One became the so-called Gauckler-Manning formula, but Robert Manning preferred the second formula. It must be noted that the Gauckler-Manning formula was proposed first by the Frenchman P.G. Gauckler (Gauckler, 1867).

McKay   Professor Gordon M. McKay (1913–1989) was a professor in civil engineering at the University of Queensland, Australia.

Meandering channel   alluvial stream characterized by a series of alternating bends (i.e. meanders) as a result of alluvial processes.

MEL culvert   see *Minimum energy loss culvert*.

Mesh (or mesh grid)   a network that is formed of cells and points for numerical simulation/calculation.

Metabolism   chemical processes occurring within living organisms in order to maintain life. For example, those causing food to be used for energy and growth.

Metric system   see *Système métrique*.

Mild slope   a channel slope is usually classified by comparing the uniform equilibrium flow depth to the critical flow depth. When the uniform equilibrium flow depth is larger than the critical flow depth, the uniform equilibrium flow is subcritical, and the slope is called a mild slope.

Minimum energy loss culvert   culvert designed with very smooth shapes to minimize energy losses. The design of a minimum energy loss culvert is associated with the concept of constant total head. The inlet and outlet must be streamlined in such a way that significant form losses are avoided (Apelt, 1983).

Momentum exchange coefficient   in turbulent flows, the apparent kinematic viscosity, or kinematic eddy viscosity, is analogous to the kinematic viscosity in laminar flows. It is called the momentum exchange coefficient, the eddy viscosity, or the eddy coefficient. The momentum exchange coefficient is proportional to the ratio of shear stress to strain rate and was first introduced by the Frenchman J.V. Boussinesq (1877, 1896).

Moody diagram   a graph presented in dimensionless form, from which the relationship between Darcy-Weisbach friction factor, Reynolds number, and surface roughness can be drawn.

Morton number   named after Rose Morton, a dimensionless number to describe the shape of bubbles or drops moving in a surrounding fluid or continuous phase.

Navier   Louis Marie Henri Navier (1785–1835) was a French engineer who primarily designed bridges but also extended Euler's equations of motion (Navier, 1823).

Navier-Stokes equation   momentum equation applied to a small control volume of incompressible fluid. It is usually written in vector notation. The equation was first derived by L. Navier in 1822 and S.D. Poisson in 1829 by a different method. It was derived later in a more modern manner by A.J.C. Barré de Saint-Venant in 1843 and G.G. Stokes in 1845.

Nomograph   abaque for graphical calculations; design chart.

Nonuniform equilibrium flow   the velocity vector varies from place to place at any instant: steady nonuniform flow (e.g. flow through an expanding tube at a constant rate) and unsteady nonuniform flow (e.g. flow through an expanding tube at an increasing flow rate).

Normal depth   uniform equilibrium open channel flow depth.

Obvert   roof of the barrel of a culvert. Another name is soffit.

One-dimensional flow   neglects the variations and changes in velocity and pressure transverse to the main flow direction. An example of one-dimensional flow can be the flow through a pipe.

One-dimensional model   model defined with one spatial coordinate, with the variables being averaged in the other two directions.

Ostraciiform fish propulsion   propulsion mode of unstreamlined and encased fish in a body armor. The main means of propulsion is by fin undulations (Lighthill, 1975; Blake, 1983). Named after the boxfish (*Ostracion*).

Outflow   downstream flow.

Outlet   (1) downstream opening of a pipe, culvert, or canal; (2) artificial or natural escape channel.

Outlet control   in a culvert, outlet control flow conditions imply that the culvert flow is controlled at the outlet (i.e. by the tailwater conditions).

Overlay zone   fill or ballast directly over the culvert, typically more than 0.15 m.

Pascal   Blaise Pascal (1623–1662) was a French mathematician, physicist, and philosopher. He developed the modern theory of probability. Between 1646 and 1648, he formulated the concept of pressure and showed that the pressure in a fluid is transmitted through the fluid in all directions.

Pascal   unit of pressure named after the Frenchman B. Pascal: one pascal equals a newton per square-meter.

Perched outlet   a perched outlet is basically an excessive vertical drop at the culvert exit.

Percolation   another name for seepage.

Piping   progressive removal of soil particles by percolating water, leading to the development of preferential flow and channels.

Pitot   Henri Pitot (1695–1771) was a French mathematician, astronomer, and hydraulician. He was a member of the French Académie des Sciences from 1724. He invented the Pitot tube to measure flow velocity in the Seine river (first presentation in 1732 at the Académie des Sciences de Paris).

Pitot tube   device to measure flow velocity. The original Pitot tube consisted of two tubes, one with an opening facing the flow. L. Prandtl developed an improved design (Howe, 1949) that provides the total head, piezometric head, and velocity measurements   it is called a Prandtl-Pitot tube.

Potamodromous fish   migratory fish species which migrate only in fresh water (e.g. bony bream, golden perch).

Potential flow   ideal fluid flow with irrotational motion.

Prandtl   Ludwig Prandtl (1875–1953) was a German physicist and aerodynamicist who introduced the concept of the boundary layer (Prandtl, 1904) and developed the turbulent "mixing-length" theory. He was a professor at the University of Göttingen.

Preissmann   Alexandre Preissmann (1916–1990) was born and educated in Switzerland. From 1958, he worked on the development of hydraulic mathematical models at SOGREAH in Grenoble (France).

Pressure outlet   pressure outlet boundary conditions require the specification of a static (gauge) pressure at the outlet boundary.

Prismatic   a prismatic channel has an unique cross-sectional shape independent of the longitudinal distance along the flow direction. For example, a rectangular channel of constant width is prismatic.

Rapidly varied flow   characterized by large changes over a short distance (e.g. sluice gate, hydraulic jump).

Red muscle   fish muscle block consisting of a relatively small quantity of fibers, designed for low-speed sustained cruising (Blake, 1983).

Reech   Ferdinand Reech (1805–1880) was a French naval instructor who first proposed the Reech-Froude number in 1852 for the testing of model ships and propellers.

Rehbock   Theodor Rehbock (1864–1950) was a German hydraulician and professor at the Technical University of Karlsruhe. His contribution to the design of hydraulic structures and physical modeling is important.

Residual   the numerical error in results.

Reynolds   Osborne Reynolds (1842–1912) was a British physicist and mathematician who first expressed the Reynolds number (Reynolds, 1883) and later the Reynolds stress (i.e. turbulent shear stress).

Reynolds number   dimensionless number proportional to the ratio of the inertial force over the viscous force.

Rheotaxis   form of taxis seen in many aquatic organisms, including fish, whereby the organism will turn toward an incoming current (positive rheotaxis).

Right abutment   abutment on the right-hand side of an observer when looking downstream.

Right bank (right wall)   looking downstream, the right bank or the right channel wall is on the right.

Roller   in hydraulics, large-scale turbulent eddy (e.g. the roller of a hydraulic jump).

SAF   St. Anthony's Falls hydraulic laboratory at the University of Minnesota (USA).

Saint-Venant   see *Barré de Saint-Venant*.

Scale effect   discrepancy between model and prototype resulting when one or more dimensionless parameters have different values in the model and prototype.

Scour   removal of bed material caused by the eroding power of the flow.

Sediment   any material carried in suspension by the flow or as bed load that would settle to the bottom in the absence of fluid motion.

Sediment load   material transported by a fluid in motion.

Sediment transport   transport of material by a fluid in motion.

Seepage   slow movement of gravitational water through soil.

Separation   in a boundary layer, a deceleration of fluid particles leading to a reversed flow within the boundary layer. The decelerated fluid particles are forced outwards, and the boundary layer is separated from the wall. At the point of separation, the velocity gradient normal to the wall is zero:

$$\left(\frac{\partial V_x}{\partial z}\right)_{z=0} = 0$$

Separation point   in a boundary layer, intersection of the solid boundary with the streamline dividing the separation zone and the deflected outer flow. The separation point is a stagnation point.

Sequent depth   in open channel flow, the solution of the momentum equation at a transition between supercritical and subcritical flow gives two flow depths (upstream and downstream flow depths). They are called sequent depths.

Similitude   correspondence between the behavior of a model and that of its prototype, with or without geometric similarity. The correspondence is usually limited by scale effects.

Siphon   pipe system discharging waters between two reservoirs or above a dam in which the water pressure becomes subatmospheric. The shape of a simple siphon is close to an omega (i.e. Ω-shape). Inverted siphons carry waters between two reservoirs with pressures larger than atmospheric. Their design follows approximately a U-shape. Inverted siphons were commonly used by the Romans along their aqueducts to cross valleys.

Slope   (1) side of a hill; (2) inclined face of a canal (e.g. trapezoidal channel); (3) inclination of the channel bottom from the horizontal.

Sluice gate   underflow gate with a vertical sharp edge for stopping or regulating flow.

Soffit   roof of the barrel of a culvert. Another name is obvert.

Specific energy   quantity proportional to the energy per unit mass, measured with the channel bottom as the elevation datum and expressed in meters of water. The specific is linked to the total head as $E = H - z_o$, where z_o is the bed elevation. The concept of specific energy, first developed by B.A. Bakhmeteff (1912), is commonly used in open channel flows.

Spillway   opening built into a dam or the side of a reservoir to release (spill) excess flood waters.

Splitter   obstacle (e.g. concrete block, fin) installed on a chute to split the flow and increase the energy dissipation.

Stage-discharge curve   relationship between discharge and free-surface elevation at a given location along a stream.

Stagnation point   the point where the velocity is zero. When a streamline intersects itself, the intersection is a stagnation point. For irrotational flow, a streamline intersects itself at a right angle at a stagnation point.

Standard k-ε model   a common turbulence model used in computational fluid dynamics (CFD) to simulate mean flow characteristics for turbulent flow conditions. It is a two-equation model that gives a general description of turbulence by means of two transport equations (PDEs). The original impetus for the k-ε model was to improve the mixing-length model, as well as to find an alternative to algebraically prescribing turbulent length scales in moderate to high complexity flows: The first transported variable is the turbulence kinetic energy (k). The second transported variable is the rate of dissipation of turbulence energy (ε).

Steady flow   occurs when conditions at any point of the fluid do not change with time:

$$\frac{\partial V}{\partial t} = 0 \ \text{ and } \ \frac{\partial P}{\partial t} = 0$$

Steep slope   when the uniform equilibrium flow depth is smaller than the critical flow depth, the uniform equilibrium flow is supercritical, and the slope is called a steep slope. The notion of steep and mild slope is not only a function of the bed slope but is also a function of the flow resistance and implicitly of the flow rate and channel roughness.

Stilling basin   structure for dissipating the energy of the flow downstream of a spillway, outlet work, chute, or canal structure. In many cases, a hydraulic jump is used as the energy dissipator within the stilling basin.

Storm water   excess water running off the surface of a drainage area during and immediately following a period of rain. In urban areas, waters drained off a catchment area during or after a heavy rainfall are usually conveyed in manmade storm waterways.

Storm waterway   channel built for carrying storm waters.

Streamline   the line drawn so that the velocity vector is always tangential to it (i.e. no flow across a streamline). When the streamlines converge, the velocity increases. The concept of streamline was first introduced by the Frenchman J.C. de Borda.

Stream tube   a filament of fluid bounded by streamlines.

Subcritical flow   in an open channel the flow is defined as subcritical if the flow depth is larger than the critical flow depth. In practice, subcritical flows are controlled by the downstream flow conditions.

Supercritical flow   in an open channel, when the flow depth is less than the critical flow depth, the flow is supercritical and the Froude number is larger than 1. Supercritical flows are controlled from upstream.

Suspended load   transported sediment material maintained into suspension.

Swimming speed   types of fish swimming speed performance are generally based on the duration of swimming to when a fish becomes fatigued and requires rest: endurance speed, also called sustained speed, prolonged speed, and burst or darting speed.

Système international d'unités   international system of units adopted in 1960 based on the meter-kilogram-second (MKS) system. It is commonly called the SI unit system. The basic seven units are for length, the meter; for mass, the kilogram; for time, the second; for electric current, the ampere; for luminous intensity, the candela; for amount of substance, the mole; and for thermodynamic temperature, the kelvin.

Système métrique   international decimal system of weights and measures that was adopted in 1795 during the French Revolution. Between 1791 and 1795, the Académie des Sciences de Paris prepared a logical system of units based on the meter for length and the kilogram for mass. The standard meter was defined as 1×10^{-7} times a meridional quadrant of earth. The gram was equal to the mass of 1 cm^3 of pure water at the temperature of its maximum density (i.e. 4 degrees Celsius) and 1 kilogram equaled 1000 grams. The liter was defined as the volume occupied by a cube of 1×10^3 cm^3.

Tailwater   downstream flow.

Tailwater depth   downstream flow depth.

Tailwater level   downstream free-surface elevation.

Tainter gate   type of radial gate, named after the American engineer J.B. Tainter.

Total head   the total head is proportional to the total energy per unit mass and per gravity unit. It is expressed in meters of water.

Training wall   sidewall of chute spillway.

Trashrack   screen comprising metal or reinforced concrete bars located at the intake of a waterway to prevent the progress of floating or submerged debris.

Turbulence   flow motion characterized by its unpredictable behavior, strong mixing properties, and broad spectrum of length scales (Lesieur, 1994).

Turbulent flow   in turbulent flows the fluid particles move in very irregular paths, causing an exchange of momentum from one portion of the fluid to another. Turbulent flows have great mixing potential and involve a wide range of eddy length scales.

Two-dimensional flow   all particles are assumed to flow in parallel planes along identical paths in each of these planes. There are no changes in flow normal to these planes. An example of two-dimensional flow can be an open channel flow in a wide rectangular channel.

Two-phase flow   flow that has more than one phase/medium (e.g. open channel flow contains two phases   one is water and one is air).

TWRC   tail water rating curve.

TWRL   tail water rating level.

Undular hydraulic jump   hydraulic jump characterized by steady, stationary, free-surface undulations downstream of the jump and by the absence of a formed roller. The undulations can extend far downstream of the jump with decaying wave lengths, and the undular jump occupies a significant length of the channel. It is usually observed for $1 < Fr_1 < 1.5$ to 3 (Chanson and Montes, 1995). The first significant study of undular jump flow can be attributed to Fawer (1937), and undular jump flows should be called Fawer jump in homage to Fawer's work.

Undular surge   positive surge characterized by a train of secondary waves (or undulations) following the surge front. Undular surges are sometimes called Boussinesq-Favre waves in homage to the contributions of J.B. Boussinesq and H. Favre.

Uniform equilibrium flow   occurs when the velocity and depth are identical at every point in terms of magnitude and direction for a given instant:

$$\frac{\partial V_{mean}}{\partial x} = 0 \text{ and } \frac{\partial d}{\partial x} = 0$$

in which time is held constant and ∂x is a longitudinal displacement. That is, steady uniform flow (e.g. liquid flow through a long pipe at a constant rate) and unsteady uniform flow (e.g. liquid flow through a long pipe at a decreasing rate).

Unsteady flow   the flow properties change with the time.

Uplift   upward pressure in the pores of a material (interstitial pressure) or on the base of a structure. Uplift pressures can lead to the destruction of stilling basins and even to the failures of concrete dams (e.g. the Malpasset dam break in 1959).

Upstream flow conditions   flow conditions measured immediately upstream of the investigated control volume. Another name is headwater conditions.

USACE   United States Army Corps of Engineers.

USBR   United States Bureau of Reclamation.

Validation   comparison between model results and prototype data to validate the model. The validation process must be conducted with prototype data that are different from that used to calibrate and verify the model.

Velocity inlet   velocity inlet boundary conditions are used to define the flow velocity, along with all relevant scalar properties of the flow, at flow inlets.

Vena contracta   minimum cross-section area of the flow (e.g. jet or nappe) discharging through an orifice, sluice gate, or weir.

Viscosity   fluid property that characterizes the fluid resistance to shear (i.e. resistance to a change in shape or movement of the surroundings).

Volume of fluid   a free-surface modeling technique in computational fluid dynamics (CFD).

Von Karman constant   see *Karman constant*.

Wake region   the separation region downstream of the streamline that separates from a boundary.

Warrie   Australian Aboriginal name for "rushing water."

Waste waterway   old name for a spillway, particularly used in irrigation with reference to the waste of waters resulting from a spill.

Wasteweir   a spillway. The name refers to the waste of hydroelectric power or irrigation water resulting from the spill. A staircase wasteweir is a stepped spillway.

Water   common name applied to the liquid state of the hydrogen–oxygen combination H2O. Although the molecular structure of water is simple, the physical and chemical properties of H2O are unusually complicated. Water is a colorless, tasteless, and odorless liquid at room temperature. One most important property of water is its ability to dissolve many other substances: H_2O is frequently called the universal solvent. Under standard atmospheric pressure, the freezing point of water is 0 degrees Celsius (273.16 K) and its boiling point is 100 degrees Celsius (373.16 K).

Weak jump   a weak hydraulic jump is characterized by a marked roller, no free-surface undulation, and low energy loss. It is usually observed after the disappearance of undular hydraulic jump with increasing upstream Froude numbers.

Weber number   dimensionless number characterizing the ratio of inertial forces over surface tension forces. It is relevant in problems with gas–liquid or liquid–liquid interfaces.

Weir   low river dam used to raise the upstream water level. Measuring weirs are built across a stream for the purpose of measuring the flow.

Wetted perimeter   considering a cross-section (elected normal to the flow streamlines), the wetted perimeter is the length of wetted contact between the flowing stream and the solid boundaries. For example, in a circular pipe flowing full, the wetted perimeter equals the circle perimeter.

Wetted surface   in an open channel, the wetted surface refers to the surface area in contact with the flowing liquid.

White muscle   fish muscle block consisting of a relatively large mass of fibers, designed for high-speed burst cruising (Blake, 1983).

Wing wall   sidewall of an inlet or outlet.

Hydraulic calculations of natural flood plain and less-than-design flow in a box culvert

B.1 Presentation

The construction of a culvert structure affects a catchment. A culvert is a covered channel designed to pass water beneath an embankment. The design can vary from a simple geometry (standard box culvert) to a hydraulically smooth shape (minimum energy loss [MEL] culvert) (Apelt, 1983; Chanson, 1999a). The following paragraphs discuss the hydraulic calculations of flood plain and of culvert operation for less-than-design flow conditions in the context of a flood plain with a mild slope.

B.2 Natural flood plain flow calculations

In the absence of structures, the river channel responds to floods in a deterministic way, set by fundamental principles – namely, the conservation of mass, momentum, and energy. In many cases, the water depth d is equal to or close to the uniform equilibrium flow depth in the flood plain for the relevant discharge Q. In other situations, basic hydraulic calculations may be conducted assuming implicitly a mild slope, for which gradually varied flow properties correspond to a subcritical flow motion and may be controlled by downstream boundary conditions (e.g. weir, riffles, change in bed slope). In this section, both uniform equilibrium and gradually varied flow calculations are presented. The results may be used to predict the tailwater conditions of a culvert structure installed in a river channel with a mild slope when river gauge data are not available.

B.2.1 Uniform equilibrium flow conditions (normal flow conditions)

For a steady uniform equilibrium open channel flow, the flow properties (i.e. depth d and mean velocity V_{mean}) are independent of time and of longitudinal position. The application of the momentum equation in an integral form yields an exact balance between the gravity force component in the flow direction and the boundary shear force. The result is a theoretical solution of the uniform equilibrium mean flow velocity:

$$V_{mean} = \sqrt{\frac{8 \times g}{f}} \times \sqrt{\frac{D_H}{4} \times \sin\theta} \qquad \text{uniform equilibrium flow} \qquad (B.1)$$

where V_{mean} is the bulk velocity, or cross-sectional averaged velocity; g is the gravity acceleration; f is the Darcy-Weisbach friction factor; D_H is the hydraulic diameter; and θ is the

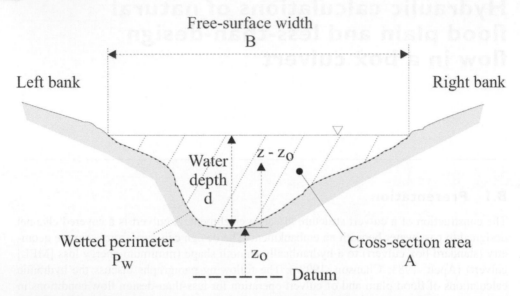

Figure B.1 Definition sketch of the flow area in a natural channel (looking downstream)

angle between the channel bed and the horizontal. The hydraulic diameter is defined as $D_H = 4 \times A/P_w$, with A the flow cross-section area[1] and P_w the wetted perimeter (Figure B.1). The term $\sin\theta$ is called the bed slope and denoted by $S_o = \sin\theta$.

Equation (B.1) is solved iteratively since the friction factor is a function of the mean flow velocity and water depth (Henderson, 1966; Chanson, 2004), while the water discharge Q must fulfill the conservation of mass:

$$Q = V_{mean} \times A \tag{B.2}$$

B.2.1.1 Discussion

In manmade channels, Equation (B.1) is the only correct expression of the momentum equation for uniform equilibrium flow in an open channel. In many practical turbulent flow situations, the Darcy-Weisbach friction factor may be estimated using the Colebrook-White formula:

$$\frac{1}{\sqrt{f}} = -2.0 \times \log_{10}\left(0.2695 \times \frac{k_s}{D_H} + \frac{2.51}{Re \times \sqrt{f}}\right) \tag{B.3}$$

where k_s is the equivalent sand roughness height of the boundary surface and Re is the Reynolds number $Re = \rho \times V_{mean} \times D_H/\mu$, with ρ the water density and μ the water dynamic viscosity.[2]

The Darcy-Weisbach friction factor appears on both sides of the Colebrook-White formula (Equation (B.3)), which must be solved iteratively. It may also be solved graphically

Table B.1 Typical equivalent sand roughness height and friction coefficients for concrete channels

Boundary surface	k_s (mm)	f	$C_{Chézy}$ ($m^{1/2}.s$)	n_{GM} ($s/m^{1/3}$)
Smooth concrete	0.3 to 3	0.012 to 0.02	62 to 80	0.012
Rough concrete	3 to 10	0.015 to 0.03	51 to 72	0.014

(e.g. using the Moody diagram). Typical values of equivalent sand roughness height and Darcy-Weisbach friction factor are reported in Table B.1.

In natural channels, the solution of the momentum equation may be rewritten in terms of the empirical friction coefficient, such as a Chézy coefficient $C_{Chézy}$ (in $m^{1/2}.s$) or Gauckler-Manning coefficient n_{GM} (in $s/m^{1/3}$):

$$V_{mean} = C_{Chézy} \times \sqrt{\frac{D_H}{4} \times \sin\theta} \qquad (B.4)$$

$$V_{mean} = \frac{1}{n_{GM}} \times \left(\frac{D_H}{4}\right)^{2/3} \times \sqrt{\sin\theta} \qquad (B.5)$$

B.2.2 Gradually varied flow conditions

For a steady gradually varied open channel flow, the differential form of the energy equation gives a relationship between the total head H and flow resistance in the form:

$$\frac{\partial H}{\partial x} = -S_f \qquad \text{gradually varied flow} \qquad (B.6)$$

where x is the longitudinal distance following the river channel and positive downstream and S_f is the friction slope defined as:

$$S_f = \frac{f}{D_H} \times \frac{V_{mean}^2}{2 \times g} \qquad (B.7)$$

The friction slope is the slope of the total head line. For an open channel flow and hydrostatic pressure distributions, the total head is:

$$H = d \times \cos\theta + z_o + \frac{V_{mean}^2}{2 \times g} \qquad (B.8)$$

with z_o the invert elevation.

Also called the backwater equation, Equation (B.6) may be applied to gradually varied steady flows in natural and manmade channels. It is valid within well-defined assumptions (Henderson, 1966; Chanson, 2004).

The backwater equation may be integrated numerically, starting from a location of a known water depth. A well-known integration method is the standard step method, distance calculated from depth, or depth calculated from distance (Montes, 1998).

B.2.2.1   Discussion

In natural channels, the friction slope might be expressed in terms of the empirical friction coefficient $C_{Chézy}$ and n_{GM}:

$$S_f = \frac{1}{C_{Chézy}^{\,2}} \times \frac{V_{mean}^{\,2}}{\dfrac{D_H}{4}} \tag{B.9}$$

$$S_f = n_{GM}^{\,2} \times \frac{V_{mean}^{\,2}}{\left(\dfrac{D_H}{4}\right)^{4/3}} \tag{B.10}$$

B.3   Hydraulic calculations of less-than-design flow in a box culvert

B.3.1   Presentation

A culvert consists of three components: the intake or inlet, the barrel or throat, and the diffuser or outlet. Current engineering practices lead to an optimum design with the smallest barrel cross-section area, with inlet control conditions for the design discharge and maximum acceptable afflux (Chanson, 2004; Concrete Pipe Association of Australasia, 2012). The following paragraphs discuss the hydraulic calculations for less-than-design flow conditions of a culvert located in a flood plain with a mild slope. For a flat flood plain and discharges substantially smaller than the design flow, the flow is subcritical in the entire culvert structure, accelerating in the inlet, fastest in the barrel, and decelerating with some energy dissipation in the outlet. In the flood plain, the flow is subcritical in the absence of a culvert structure. With the culvert structure installed in the ground level, the tailwater conditions are the same as in the absence of the culvert, and the tailwater depth is denoted d_{tw} (Fig. B.2).

B.3.2   Application

For small discharges, the flow is subcritical and best controlled from downstream (Fig. B.2). Hydraulic calculations are performed from the tailwater, where the water discharge Q and the flow depth d_{tw} are known.[3] For a low flow, outlet control takes place basically (Bates *et al.*, 2003; Hotchkiss and Frei, 2007). In the outlet, flow separation and form losses occur for expansion angles greater than 5 to 8 degrees (Montes, 1998; Chanson, 2004). In practice, a majority of culvert outlets are built with wingwalls oriented between 30 and 60 degrees from the culvert barrel centerline. The energy losses in the outlet are significant and must be estimated accurately. For a horizontal channel, the application of the energy equation between the culvert barrel exit and the tailwater flow yields:

$$d_{exit} + \frac{V_{exit}^{\,2}}{2 \times g} = d_{tw} + \frac{V_{tw}^{\,2}}{2 \times g} + K_{out} \times \left(\frac{V_{exit}^{\,2}}{2 \times g} - \frac{V_{tw}^{\,2}}{2 \times g} \right) \tag{B.11}$$

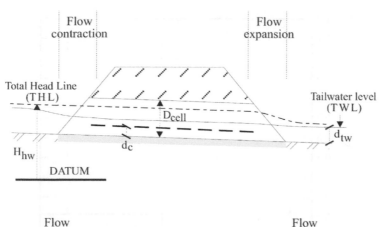

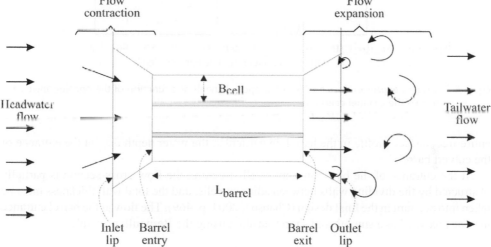

Figure B.2 Definition sketch of standard box culvert operation for less-than-design flow conditions in a flood plain with a mild slope

where d is the water depth, V is the mean velocity, the subscript exit refers to the culvert barrel exit flow conditions, the subscript tw refers to the tailwater flow conditions, and the coefficient K_{out} is an outlet loss coefficient.[4]

Experimental observations indicated that $0.8 < K_{out} < 1.1$ for a divergence angle from centerline greater than 20 degrees (Montes, 1998) (Fig. B.3). It is commonly assumed that $K_{out} = 1$ (Henderson, 1966). Combining Equation (B.11) and the equation of conservation of mass yields the flow properties at the downstream end of the culvert barrel (i.e. the depth d_{exit} and velocity V_{exit}).

Upstream of the barrel exit, the flow from the upstream flood plain into the inlet and culvert barrel is basically gradually varied. In the culvert barrel, the application of the differential form of the energy equation (i.e. the backwater equation) enables the prediction of the free-surface profile in the barrel (Section B.2.2). Its numerical integration predicts the

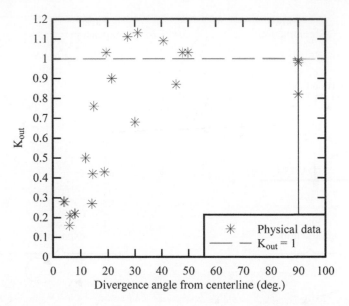

Figure B.3 Straight expansion loss coefficient in open channels as a function of the opening angle relative to the channel centerline

entire free-surface profile in the barrel, in particular, the water depth d_{entry} at the entrance of the culvert barrel.

At the entrance of the barrel in a multicell structure, the flow cross-section is partially obstructed by the dividing walls between adjacent cells, and the total wall thickness must be taken into account in the final design (Chanson, 2004, p. 469). The flow at the barrel entrance may be analyzed as a smooth and short transition using the Bernoulli principle:

$$d_{in} + \frac{V_{in}^{2}}{2 \times g} + z_{o} = d_{entry} + \frac{V_{entry}^{2}}{2 \times g} + z_{o} \tag{B.12}$$

where the subscript in refers to the downstream end of the inlet and the subscript entry corresponds to the culvert barrel entrance flow conditions.

The combination of the Bernoulli equation (Equation (B.12)) and equation of conservation of mass gives the flow properties at the downstream end of the culvert inlet.

Assuming an inlet with wingwalls,[5] backwater calculations are performed from the downstream end of the inlet to the inlet lip. Note that the backwater calculations may be performed using the standard step method depth calculated from distance.

Upstream of the inlet, the flow transition from the upstream flood plain to the start of the inlet may be analyzed using the Bernoulli principle. While the cross-section may be assumed to be rectangular at the inlet lip (in the first approximation), the upstream flood plain is a natural channel, and its properties must be used to estimate the flow cross-section area. The results give the water depth and total head in the upstream flood plain at less-than-design flow. In turn, the corresponding afflux may be calculated.

B.3.3   Discussion

This method describes one-dimensional calculations of the entire culvert flow for less-than-design conditions, assuming a mild slope; in the absence of a hydraulic jump in the outlet and immediately downstream of the outlet lip; and without obstacles (e.g. debris) in the culvert inlet, barrel, and outlet. The less-than-design flow calculations are conducted assuming free-surface flow in the inlet, barrel, and outlet, with the flow being assumed to remain subcritical in the barrel and that no hydraulic jump takes place in the outlet.

The complete free-surface profile and total head line at the embankment centerline may show some sharp change in flow properties at the start of the inlet, at the transition from inlet to barrel, and at the outlet (Fig. B.4). These changes are linked to some rapid local flow transition (e.g. flow contraction, flow expansion). Figure B.4 presents a typical example of less-than-design flow calculations for a multicell box culvert. Note that the hydraulic engineering calculations are performed from downstream to upstream, since the flow is controlled from downstream (i.e. by the tailwater flow conditions).

In the culvert barrel, the application of the differential form of the energy equation enables a prediction of the free-surface profile.

$$\frac{\partial\left(d + \dfrac{V_{mean}^{2}}{2 \times g}\right)}{\partial x} = S_{o} - S_{f} \tag{B.13}$$

where S_{o} is the bed slope ($S_{o} = \sin\theta$) and S_{f} is the friction slope defined as:

$$S_{f} = \frac{f}{D_{H}} \times \frac{V_{mean}^{2}}{2 \times g} \tag{B.14}$$

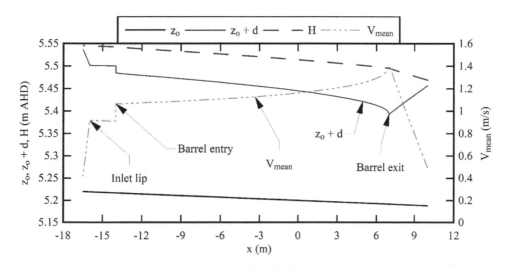

Figure B.4 Typical example of the longitudinal profile of water depth, mean flow velocity, and total head in a multicell box culvert operating at less-than-design flow with subcritical free-surface flow – Q_{des} = 4.8 m³/s, S_{o} = 0.0012, seven cells, barrel length = 14 m, B_{cell} = 1.00 m, D_{cell} = 0.500 m, Q = 1.3 m³/s, d_{tw} = 0.21 m

The barrel flow calculations are typically undertaken for smooth boundaries, and friction factor calculations are discussed in Section B.2.1. In the presence of rough barrel walls or culvert barrel baffles, the flow resistance estimates must be derived from detailed hydrodynamic calculations validated by solid engineering data.

For the culvert barrel flow calculations, great attention must be paid to the wetted perimeter estimates in a multicell structure. At the entrance of the barrel in a multicell structure, the flow cross-section is obstructed by the dividing walls, and the transition between the end of the inlet to the barrel entrance may be analyzed using the Bernoulli principle.

At the entrance of the inlet (i.e. inlet lip), the flow undergoes a contraction. The transition from the upstream flood plain to the start of the inlet may be analyzed using the Bernoulli principle.

Notes

1 The flow cross-section area is measured perpendicular to the streamlines.
2 At 20 degrees Celsius, the water density and dynamic viscosity are, respectively, $\rho = 998.2$ kg/m³ and $\mu = 0.001005$ Pa.s.
3 The relationship between water discharge and tailwater depth may be given by water gauge data or estimated based upon gradually varied flow or uniform equilibrium flow calculations applied to the downstream flood plain.
4 For $K_{out} = 1$, Equation (B.11) yields the Borda-Carnot formula for a sudden expansion.
5 A simple inlet geometry consists of 45-degree straight wingwalls.

Appendix C

On physical modeling of fish passage in standard box culverts

C.1 Presentation

Culverts are common hydraulic structures along rivers and streams in rural and urban water systems. A culvert is a covered channel allowing the passage of flood waters beneath an embankment, for example, a roadway. The designs are very diverse, using various shapes and construction materials determined by stream width, peak flows, stream gradient, road direction, and minimum cost (Hee, 1969; Australian Standard, 2010). The hydraulic modeling of culverts aims to find optimal solutions of full-scale culvert prototype structures (Novak and Cabelka, 1981). Figure C.1 present three-dimensional scale models of box culverts, and Figure C.2 shows box culvert barrel models in which upstream fish passage was tested. For the modeling of upstream fish passage in culverts, major differences between up-scaled laboratory model predictions and prototype observations may result from a lack of standardized methodology (Cotel and Webb, 2015; Katopodis and Gervais, 2016). Recent contributions hinted that "a proper study of turbulence effects on fish behaviour should involve, in addition to turbulence energetics, consideration of fish dimensions in relation to the spectrum of turbulence scales" (Nikora et al., 2003, p. 1380), as well as "the scale of the turbulence with respect [to] the fish" (Lacey et al., 2012, p. 430).

In this section, the modeling of upstream passage of fish in box culverts is reviewed, with a focus on the fish–turbulence interactions and the implications in terms of laboratory modeling and upstream fish passage, following Wang and Chanson (2018a). Dimensional analysis provides a number of important dimensionless parameters relevant to all laboratory studies. Practical limitations are discussed. Later laboratory studies of fish swimming in a full-scale box culvert barrel are presented and seminal findings are highlighted.

C.2 Physical modeling, dimensional considerations, and similitude

C.2.1 Physical modeling and dimensional considerations

In experimental fluid mechanics, a laboratory model is used to provide reliable predictions of the fluid dynamics properties in the full-scale hydraulic structure (Henderson, 1966; Foss et al., 2007). The physical modeling must be based upon the fundamental concepts and principles of similitude to ensure a sound extrapolation of the scaled model results. In the dimensional analysis of fluid mechanics, the relevant dimensional variables include the fluid properties, physical constants, boundary conditions (including channel geometry), and initial flow conditions (Liggett, 1994).

(A) 1:30 scale model of a large culvert structure beneath the Zurich main railway station in Switzerland in the VAW, ETZ-Zürich in 2013 – the modeling was undertaken to verify the accuracy of prototype culvert discharge capacity

(B) Single-cell box culvert model at the University of Queensland (Australia) for $Q/Q_{des} = 1.2$ and different tailwater depth – from top, counterclockwise: outlet operation for $d_{tw}/D = 0.62$, outlet operation for $d_{tw}/D = 1.1$, submerged inlet operation for $d_{tw}/D = 1.1$

Figure C.1 Physical modeling of standard box culvert structures

Considering a steady turbulent flow in a standard box culvert barrel operating as a free-surface flow, the dimensional analysis gives a series of relationships between the fluid dynamics characteristics at a location (x, y, z) and the upstream flow conditions, boundary conditions, and fluid properties:

$$d, \vec{V}, v', p, L_t, T_t, \ldots = F_1\left(x, y, z, B, k_s, \theta, h_b, L_b, d_1, V_1, v_1', \rho_w, \mu_w, \sigma, g, \ldots\right) \tag{C.1}$$

where d is the flow depth; V is the velocity; v' is a velocity fluctuation; p is the pressure; L_t and T_t are integral turbulent length and time scales; x, y, and z are, respectively, the longitudinal, transverse, and vertical coordinates; B is the internal barrel width; k_s is the equivalent sand roughness height of the culvert barrel boundary; θ is the angle between the culvert invert and horizontal; h_b and L_b are, respectively, the height and longitudinal spacing of simplistic baffles; d_1, V_1, and v_1' are, respectively, the inflow depth, velocity, and velocity fluctuation; ρ_w and μ_w are the water density and dynamic viscosity; σ is the surface tension; and g is the gravity acceleration.

The Π-Buckingham theorem states that any dimensional equation with N variables with units encompassing mass, length, and time may be rewritten into an equation with (N − 3) dimensionless parameters (Vaschy, 1892; Buckingham, 1914; Rouse, 1938). It is also called Vaschy-Buckingham theorem after the French engineer Aimé Vaschy (1857–1899) and American physicist Edgar Buckingham (1867–1940). In turn, Equation (C.1) may be expressed:

$$\frac{d}{d_c}, \frac{V_x}{V_c}, \frac{v_x'}{V_c}, \frac{p}{\rho_w \times g \times d_c}, \frac{L_t}{d_c}, T_t \times \sqrt{\frac{g}{d_c}} \ldots = F_2 \left(\begin{array}{c} \dfrac{x}{d_c}, \dfrac{y}{d_c}, \dfrac{z}{d_c}, \\[2mm] \dfrac{B}{d_c}, \theta, \dfrac{k_s}{d_c}, \dfrac{h_b}{d_c}, \dfrac{L_b}{d_c}, \\[2mm] \dfrac{d_1}{d_c}, \dfrac{V_1}{\sqrt{g \times d_1}}, \dfrac{v_1'}{V_1}, \\[2mm] \rho_w \times \dfrac{V \times D_H}{\mu_w}, \dfrac{g \times \mu_w^4}{\rho_w \times \sigma^3}, \ldots \end{array} \right) \tag{C.2}$$

where d_c is the critical flow depth: $d_c = (Q^2/(g \times B^2))^{1/3}$, V_c is the critical flow velocity, Q is the water discharge, and D_H is the equivalent pipe diameter, or hydraulic diameter. In Equation (C.2), the seventh term in the term on the right is the inflow Froude number Fr_1, and the eighth and ninth terms are the Reynolds number Re and Morton number Mo, respectively. Note that the Morton number is introduced because it becomes a constant in most hydraulic model studies, when air and water are used in both laboratory experiments and prototype flows (Kobus, 1984; Chanson, 2009b).

Traditionally, hydraulic laboratory modeling is conducted using a geometrically similar model (Chanson, 1999b, 2004). Geometric similarity implies that the ratios of prototype characteristic lengths to model lengths are equal. If any form of similarity (i.e. geometric, kinematic, or dynamic) is not satisfied, scale effects may take place, yielding substantial differences between the laboratory data extrapolation and the culvert prototype structure performances. In a physical model, true similarity can be achieved if and only if all dimensionless parameters or Π-terms have the same values in both laboratory and full-scale structure:

$$
\begin{aligned}
\mathrm{Fr_m} &= \mathrm{Fr_p} \\
\mathrm{Re_m} &= \mathrm{Re_p} \\
\mathrm{Mo_m} &= \mathrm{Mo_p}
\end{aligned}
\tag{C.3}
$$

where the subscripts m and p refer to the laboratory model and full-scale conditions, respectively. Open channel flows, including culvert flows, are traditionally investigated based upon a Froude similarity because gravity effects are important (Henderson, 1966; Novak and Cabelka, 1981). When the same fluids – air and water – are used in the laboratory and at full scale, the Froude and Morton similarities are applied simultaneously. Then the Reynolds number may be grossly underestimated in small laboratory flumes (Figs. C.1B, C.2B, and C.2C).

A dimensional analysis must be similarly undertaken for the fish motion in a turbulent culvert barrel flow (Alexander, 1982; Wang and Chanson, 2018a). Considering the upstream passage of a fish in a prismatic box culvert barrel with steady turbulent flow conditions, a dimensional analysis gives a series of relationships between the fish motion at a given location (x, y, z), fish properties (including species), channel boundary conditions, turbulent flow properties, fluid properties, and physical constants:

$$
\vec{U}, u', O_2, \tau_f, \ldots = F_3
\begin{pmatrix}
x, y, z, \\
L_f, l_f, h_f, \rho_f, \text{specie}, \\
B, k_s, \theta, \\
d, V, v', L_t, T_t, \\
\rho_w, \mu_w, \sigma, g, \ldots
\end{pmatrix}
\tag{C.4}
$$

where U is the fish speed positive upstream; u' is a fish speed fluctuation; O_2 is the oxygen consumption; τ_f is the fish response time; L_f, l_f, and h_f are, respectively, the fish length, thickness, and height; and ρ_f is the fish density. Herein Equation (C.4) ignores the effects of fish fatigue, heat transfer, and metabolism. The application of the Π-Buckingham theorem shows that Equation (C.4) may be transformed into a dimensionless form:

$$
\frac{U}{V_c}, \frac{u'}{v'}, O_2, \frac{\tau_f}{T_t}, \ldots = F_4
\begin{pmatrix}
\dfrac{x}{d_c}, \dfrac{y}{d_c}, \dfrac{z}{d_c}, \\
\dfrac{L_f}{L_t}, \dfrac{l_f}{L_t}, \dfrac{h_f}{L_t}, \dfrac{\rho_f}{\rho_w}, \text{specie}, \\
\dfrac{B}{d_c}, \dfrac{k_s}{d_c}, \theta, \\
\mathrm{Fr}, \mathrm{Re}, \mathrm{Mo}, \dfrac{L_t}{d_c}, T_t \times \sqrt{\dfrac{g}{d_c}}, \ldots
\end{pmatrix}
\tag{C.5}
$$

Equation (C.5) demonstrates several key dimensionless variables most relevant to the upstream passage of fish in a turbulent culvert barrel flow. Such fundamental parameters encompass the ratio u'/v' of fish speed fluctuations to fluid velocity fluctuations, ratio τ_f/T_t of fish response time to turbulent time scales, ratios of fish dimension to turbulent length scale, and the fish species (Wang and Chanson, 2018a). Considering the upstream fish passage in a turbulent culvert flow, the extrapolation of the laboratory model data to the full-scale culvert will be achievable only if all the relevant key dimensionless parameters shown in Equation (C.5) are the same in the laboratory and in the full-scale culvert structure.

A few studies recorded quantitative detailed characteristics of both fish motion and fluid flow (Nikora *et al.*, 2003; Plew *et al.*, 2007; Wang *et al.*, 2016a). Fewer investigations reported fish speed fluctuations and fluid velocity fluctuations, as well as fish response time and integral time scales (Wang *et al.*, 2016a; Cabonce *et al.*, 2018). All the results showed that a number of key parameters, including the ratios u'/v', τ_f/T_t, and L_f/L_t, are scale dependent when the same fish are used in the laboratory and in the field, as shown by Equation (C.5). In other words, a complete similarity between laboratory data and full-scale observations may be unachievable in small-size flumes (e.g. Figs. C.1, C.2B, and C.2C). One must seek either some near-full-scale testing (Fig. C.2A), an incomplete similitude, some approximate estimate, or an alternative approach.

(A) 12 m long and 0.5 m wide rectangular channel, with flow direction from background to foreground – experimental low conditions: $\theta = 0$, $Q = 0.0556$ m³/s, $d = 0.166$ m, $Re = 1.4 \times 10^5$, small triangular baffles on left bottom corner, looking upstream

Figure C.2 Laboratory studies of upstream fish passage in box culvert barrel

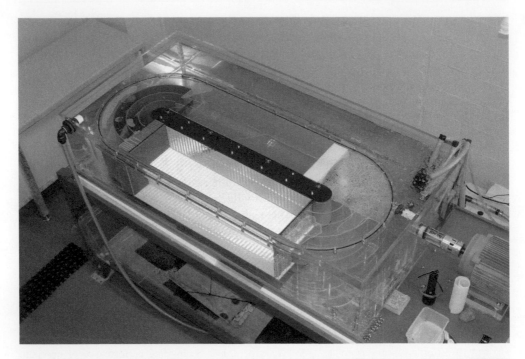

(B) Recirculation water tunnel – test section: B = 0.25 m, d = 0.26 m, = 0, V ~ 0.05 m/s, counterclockwise flow recirculation

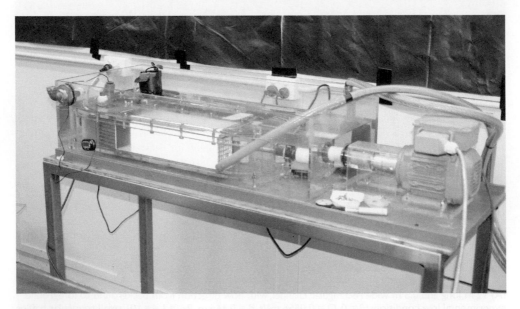

(C) Small recirculation water tunnel – test section: B = 0.10 m, d = 0.10 m, θ = 0

Figure C.2 (Continued)

C.2.2   Discussion

Most open channel flow structures, including culverts, are scaled based upon Froude and Morton similarities, when the same fluids – air and water – are used in the laboratory and prototype. That is, the Froude and Morton numbers are identical in the physical model and at full scale. Then the Reynolds number Re may be grossly underestimated in small-size flumes: for example, with Reynolds numbers of about 4.4×10^5, 5×10^3, and 2×10^3 in the laboratory models seen in Figures C.2A, C.2B, and C.2C, respectively, compared to a full-scale culvert barrel flow corresponding to Re $\approx 5 \times 10^5$ to 1×10^8 at design discharges.

The inflow conditions must be similarly scaled between full-scale culverts and laboratory flumes. Figure C.3 presents velocity contours V_x/U_{mean} based upon detailed velocity measurements in the experimental channels shown in Figures C.2A and C.2B. The flow conditions are summarized in the figure caption and in Table C.1, where D is the internal height or flow depth, B is the internal width, D_H is the equivalent pipe diameter, Re is the Reynolds number defined as Re = $\rho \times V_{mean} \times D_H / \mu$, V_{mean} is the cross-sectional averaged velocity (also called bulk velocity), and V_{75} and V_{25} are, respectively, the third and first quartiles of the cross-section velocity data set. For a Gaussian distribution of the data set about its mean, the difference between the third and first quartile would be equal to 1.3 times the standard deviation (Spiegel, 1972). The same contour color scale was used for all plots. The results illustrate the nonuniformity of the inflow conditions in the small recirculation water tunnel, also discussed by Kern et al. (2018). Results obtained in such small water tunnels are unlikely to be applicable to full-size culvert structures without significant bias and errors.

A related challenge is the physical modeling of upstream fish passage in culverts and the associated limitations and significance of current fish swim tunnel tests (Katopodis and Gervais, 2016). One may question the relevance of water tunnel testing, for example, in facilities seen in Figures C.2B and C.2C, to upstream fish passage in culverts, when field observations reported fish seeking low-velocity zones, associated with high turbulence intensity levels, to pass through hydraulic structures (Behlke et al., 1991; Blank, 2008; Goettel et al., 2015; Cabonce et al., 2018, 2019). The hydrodynamic conditions in full-size culverts differ substantially from water tunnel and tube testing conditions.

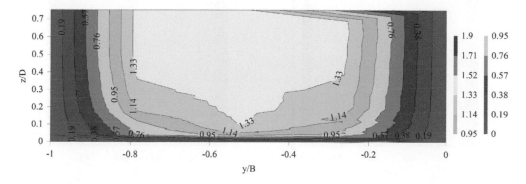

(A) 12 m long and 0.5 m wide smooth rectangular channel (see Fig. C.2A)

Figure C.3 Dimensionless contour plots V_x/V_{mean} as a function of y/B and z/D in smooth rectangular flumes – the same contour color scale was used for all plots, and the legend corresponds to V_x/V_{mean} – photographs of the laboratory flumes are shown in Figures C.2A and C.2B

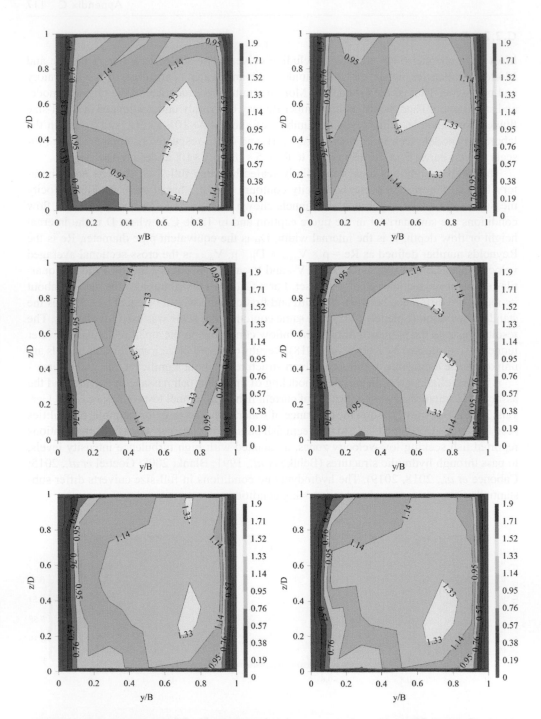

(B) 0.25 m wide and 0.26 m high smooth rectangular water tunnel (see Fig. C.2B) – from left to right, top to bottom, settings = 6, 9, 12, 15, 18, and 21

Figure C.3 (Continued)

Table C.1 Dimensional and dimensionless parameters for the 12 m long flume and recirculating water tunnel

Flume	Settings	D (m)	B (m)	D_H	Re	V_{mean} (m/s)	V_{max}/V_{mean}	$(V_{75}\text{-}V_{25})/U_{mean}$
12 m long flume	Q = 0.026 m³/s	0.154	0.5	0.375	1.3×10^5	0.35	1.40	0.40
Recirculating water tunnel	6	0.259	0.251	0.244	0.73×10^5	0.30	1.50	0.30
	9				0.92×10^5	0.38	1.42	0.24
	12				1.3×10^5	0.53	1.47	0.34
	15				1.5×10^5	0.63	1.47	0.20
	18				1.8×10^5	0.73	1.39	0.15
	21				2.1×10^5	0.85	1.45	0.18

C.3   Laboratory studies of fish swimming in a full-scale box culvert barrel

C.3.1   Presentation

Open channels and culvert barrel flows are modeled based upon a Froude similarity because gravity effects are important in terms of the hydrodynamics (Henderson, 1966). Viscous-scale effects are likely to be experienced in very small-size models, water tunnels, and water tubes, and the results cannot be extrapolated to a full-scale culvert without major bias (i.e. scale effects). When the hydrodynamics and fish kinematics are considered together, the similitude requirements become impossible to fulfill unless full-scale studies are undertaken. Only measurements in a real culvert and full-scale laboratory experiments are appropriate.

While the use of large channels to test fish swimming performance is not new (Colavecchia et al., 1998; Guiny et al., 2005; Haro et al., 2004; Richmond et al., 2007; Khodier and Tullis, 2014), this approach was recently undertaken under controlled flow conditions in a full-scale 12 m long and 0.5 m wide rectangular culvert barrel flume (Wang et al., 2016a, 2016b; Cabonce et al., 2017, 2018, 2019) (Fig. C.4). The facility would correspond to a single-cell (B = 0.5 m) box culvert beneath a typical two-lane road embankment. Both detailed hydrodynamic measurements and high-speed kinematic observations were conducted. Basic experiments with small-body-mass fish were undertaken with a range of boundary configurations (Table C.2), namely, a smooth rectangular channel, a channel with a rough bed, a channel with a rough bed and rough left sidewall,[1] and a smooth flume with small triangular corner baffles (Figs. C.4C and C.4D). The boundary roughness and small baffles were selected to have dimensions comparable to the fish dimensions, because fish benefit from large-scale turbulence when the eddy size is comparable to the fish size, and the baffle spacing was selected to create a wake–interference turbulent flow regime most beneficial to upstream fish passage. The experimental flow conditions are summarized in Table C.2, including the boundary conditions, water discharge, bulk velocity, fish sample size, fish data, and water temperature. During each series of tests, the flow rate was maintained constant, irrespective of the boundary treatment and appurtenance. The methodology enabled a direct comparison of fish swimming performances and behavior in a culvert barrel between different types of boundary treatment and delivered biological data compatible with

(A) Adult Duboulay's rainbowfish (*Melanotaenia duboulayi*) in a smooth channel, Q = 0.0261 m³/s, d = 0.123 m, Re = 2.1 × 10⁵

(B) Juvenile silver perch (*Bidyanus bidyanus*) in a channel with a rough bed and left sidewall, Q = 0.0261 m³/s, d = 0.129 m, Re = 2.2 × 10⁵

Figure C.4 Laboratory study of culvert fish passage in 12 m long and 0.5 m wide culvert barrel flume for less-then-design discharges (Table C.2), with flow direction from left to right

(C, Left) Juvenile silver perch (*Bidyanus bidyanus*) with a flume with smooth boundaries and small triangular corner baffles: h_b = 0.067 m, L_b = 0.67 m, Q = 0.0556 m³/s, d = 0.162 m, Re = 4.4 × 10⁵ – the fish was swimming upstream of a baffle, using the local stagnation region to minimize its energy expenditure

(D, Right) Juvenile silver perch (*Bidyanus bidyanus*) with a flume with smooth boundaries and small triangular corner baffles: h_b = 0.067 m, L_b = 0.67 m, Q = 0.0556 m³/s, d = 0.162 m, Re = 4.4 × 10⁵ – the fish was swimming along the left side-wall immediately downstream of a baffle and was facing downstream

Figure C.4 (Continued)

engineering design procedures and applicable by professional engineers. A review of the fish behavior and detailed hydrodynamic results follows, and Appendix F presents a series of high-speed video movies of fish swimming in the presence of different channel boundary types.

C.3.2 Basic results

Fish endurance tests were conducted with two discharges for up to 20 minutes with five boundary conditions (Table C.2). For each series, the water discharge was kept identical and the impact of boundary treatments on fish passage was tested while the flow rate remained constant for the results to be compatible with engineering design procedures and usable by

Table C.2 Review of laboratory studies on fish swimming in a 12 m long and 0.5 m wide culvert barrel model: fish data (mass m_f and total length L_f)

Reference	Q (m³/s)	d (m)	V_{mean} (m/s)	T (°C)	Fish species	Nb of fish	Fish mass m_f (g)	Fish length L_f (mm)
(1)	(2)	(3)	(4)	(5)	(6)	(7)	(8)	(9)
Wang et al. (2016a)								
Smooth channel	0.0261	0.123	0.424	24.5	Duboulay's rainbowfish (Melanotaenia duboulayi)	22	2.75 ± 0.65	68.5 ± 6.3
Rough bed and smooth sidewalls	0.0261	0.133	0.392	±0.5	Duboulay's rainbowfish (Melanotaenia duboulayi)	23	3.6 ± 1.08	74.0 ± 5.5
Rough bed and rough left sidewall	0.0261	0.129	0.424		Juvenile silver perch (Bidyanus bidyanus)	23	39.7 ± 33.7	145 ± 31.5
					Duboulay's rainbowfish (Melanotaenia duboulayi)	23	3.2 ± 1.07	70.5 ± 8.0
Cabonce et al. (2018, 2019)								
Smooth channel	0.0556	0.162	0.686	24.5	Juvenile silver perch (Bidyanus bidyanus)	20	1.50 ± 1.16	53.0 ± 11.8
Medium baffles (h_b = 0.067 m)	0.0556	0.1625	0.684	±0.5		26	1.30 ± 0.85	47.0 ± 9.6
Large baffles (h_b = 0.133 m)	0.0556	0.173	0.643			26	3.70 ± 2.81	70.5 ± 16.7
Large baffles (h_b = 0.133 m) with perforation ($\varnothing$ = 13 mm)	0.0556	0.173	0.643			15	3.20 ± 1.40	66.0 ± 8.7

Notes: d: water depth; h_b: isosceles triangular baffle size; Q: water discharge; T: water temperature; V_{mean}: bulk velocity; fish data: median value ± standard deviation; 1 values recorded 8 m downstream of the flume's entrance.

professional engineers. Figure C.5 regroups the cumulative percentages of fish swimming after test durations ranging from 1 to up to 20 minutes for two different series of tests with different discharges, boundary conditions, and fish species (Table C.2). For the larger discharge, a sizable number of fish fatigued before the end of testing – 12 out of 20 – in the smooth boundary flume (Fig. C.5B). The observations showed overall that the presence of small triangular

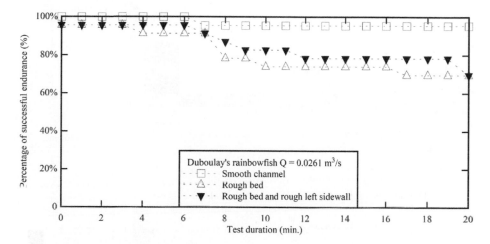

(A) Duboulay's rainbowfish (*Melanotaenia duboulayi*) in a smooth and rough channel for a relatively small flow: Q = 0.0261 m³/s, S₀ = 0

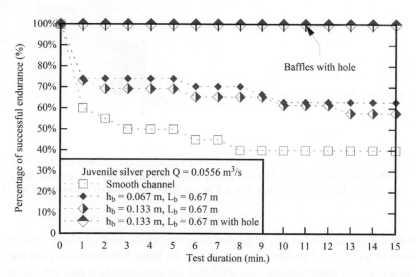

(B) Juvenile silver perch (*Bidyanus bidyanus*) in a smooth flume with small triangular corner baffles along the lower-left corner for a relatively large discharge: Q = 0.0556 m³/s, S₀ = 0

Figure C.5 Cumulative endurance test duration data for small-body-mass fish negotiating upstream passage in a 12 m long and 0.5 m wide culvert barrel flume for less-than-design discharges (Table C.2) – comparison between different boundary treatments

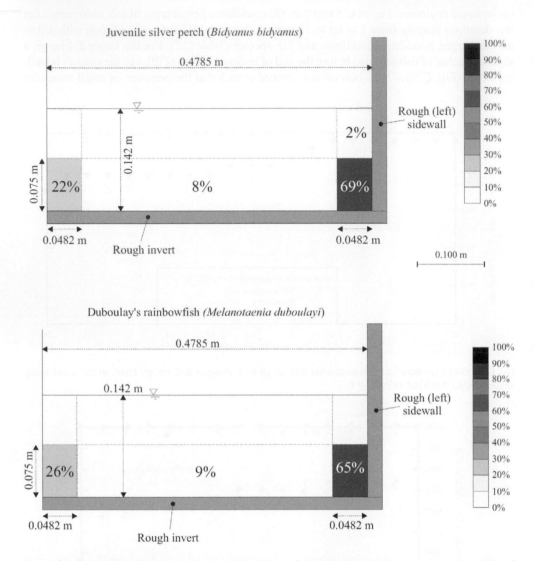

Figure C.6 Percentage of time spent by small-body-mass fish within the channel cross-section, weighted with respect to time in a channel with a rough bed and rough left sidewall for a relatively small flow: Q = 0.0261 m³/s, x = 4–6.5 m, y = 0 at right smooth sidewall (on left of graph)

corner baffles allowed fish to rest and facilitated substantially their upstream passage, including in terms of quantitative endurance swim results, for these flow conditions and fish species.

The fish position observations indicated that the small-body-mass fish swam against the current, mostly next to the bottom corners and along the sidewall of the channel (Figs. C.6 and C.7). Visual recordings are reported in Figures C.6 and C.7 for two species and five boundary treatments. All the results indicated that the small fish swam in the bottom corners and along the sidewalls for more than 90% of the time. The findings were consistent with earlier studies with small-bodied fish (Gardner, 2006; Jensen,

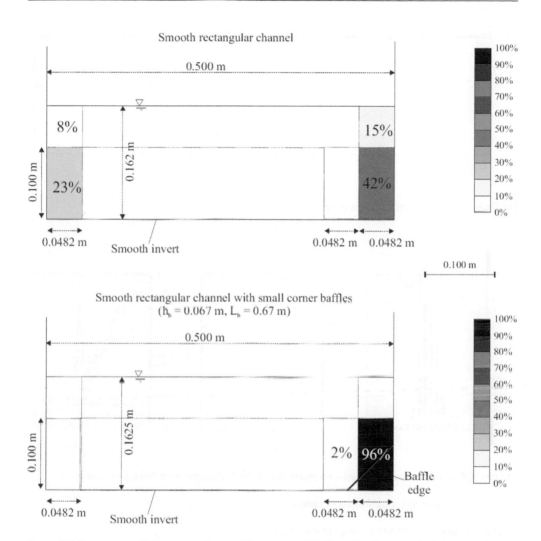

Figure C.7 Percentage of time spent by small-body-mass fish (juvenile silver perch *Bidyanus bidyanus*) within the channel cross-section, weighted with respect to time in a smooth channel (top) and channels with small triangular baffles in the bottom-left corner (lower three sketches) for a moderate discharge: Q = 0.0556 m³/s, x = 4–6.5 m, y = 0 at right smooth sidewall (on left of graph)

2014) and confirmed in very recent work (Duguay *et al.*, 2018). Visual observations, fish trajectory data, and speed time series indicated that for both fish species, the time series could be subdivided into (a) quasi-stationary motion where fish speed fluctuations were small, (b) short upstream motion facilitated by a few strong tailbeats, and (c) burst swimming when the fish would cross rapidly the observation window. The most common observation of fish swimming was the first one (i.e. quasi-stationary motion with small

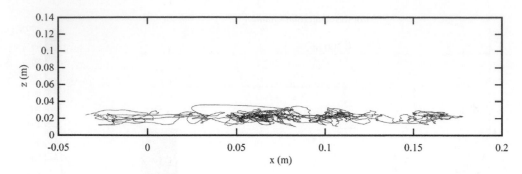

(A) Trajectory in the vertical plane (x,z)

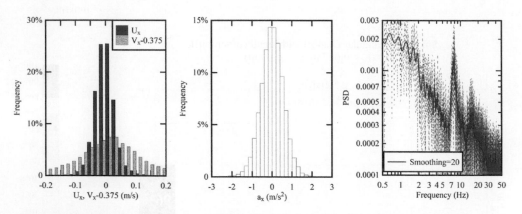

(B, Left) PDF of longitudinal fish speed – comparison with fluid velocity PDF at fish tracking location

(C, Right) PDF of longitudinal fish acceleration

(D) Power spectrum density of longitudinal fish speed

Figure C.8 Fish kinematics of an individual Duboulay's rainbowfish (*Melanotaenia duboulayi*) swimming along the rough sidewall in a channel with a rough bed and rough left sidewall for a relatively small flow: Q = 0.0261 m³/s

fish speed fluctuations). See, for example, the movies CIMG1655.mov, CIMG2672.mov, CIMG1647.mov, CIMG2725.mov, and CIMG1651.mov in a smooth box culvert channel (Appendix F).

Fish trajectories, time series of fish trajectories, fish speed, and acceleration data were recorded for all boundary conditions. Typical data for a small fish individual are presented in Figure C.8, where x is positive downstream, z is positive upwards with z = 0 at the bed, and the Eulerian fish speed U_x and acceleration a_x are positive downstream. Figure C.8A shows the fish trajectory in the (x,z) plane close to the rough sidewall (right corner in Fig. C.5).

Figures C.5B and C.5C present the probability density functions (PDFs) of fish speed and acceleration, and Figure C.5D shows the results of a spectral analysis of the fish speed time series.

Overall, the observations indicated that the fish speed was within ± 0.2 m/s and the longitudinal fish acceleration was within ± 2 m/s² (i.e. ± 0.2 × g with g the gravity acceleration). A typical example of longitudinal fish speed and acceleration data set is shown in Figure C.8 for an individual swimming in the bottom-left corner. The swimming speed variability may be compared to the distribution of longitudinal fluid velocity component at the location where the fish was tracked. In Figure C.5B, the PDF of fish speed (red bars) is plotted together with the PDF of the fluid velocity (black bars). For the small fish species, the ratio of fish speed to fluid velocity standard deviations was typically within $0.1 < u_x'/v_x' < 1$ with a median value about 0.25, independently of the fish species, total length, and mass, and v_x' being the velocity fluctuations at the observed fish location. The results are reported in Figure C.9A, presenting the ratio of fish speed to fluid velocity standard deviations as a function of the dimensionless fish length L_f/k_s with k_s the equivalent sand roughness height of the channel. Physically, the equivalent sand roughness height k_s characterizes the rugosity of the wall and is related to the size of vortical structures in the vicinity of the wall (Hong et al., 2011). Thus, the ratio L_f/k_s provides some indication of the ratio of fish length to turbulence length scale next to the wall. Herein the fish speed fluctuations were mostly smaller than the fluid velocity fluctuations, with a median value $u_x'/v_x' \sim 0.3$ for juvenile silver perch and $u_x'/v_x' \sim 0.1$ for adult Duboulay's rainbowfish. It is thought that swimming in the channel corner may allow fish to minimize the energy costs associated with changes in acceleration (Nikora et al., 2003; Wang and Chanson 2018a, 2018b).

The ratio of fish speed to fluid velocity auto-correlation time scales was typically within $0.03 < t_{xx}/T_{xx} < 5$ with a median value about 1.5 (Fig. C.9B). The fish speed auto-correlation time scale gives some information on the reaction time of the fish, while the fluid velocity auto-correlation time scale, also called the Eulerian integral turbulent time scale, is a rough measure of the longest connection in the turbulent behavior (O'Neill et al., 2004; Chanson, 2009b). The ratio t_{xx}/T_{xx} basically provides some information on the fish response time relative to the characteristic time scale of large turbulent structures. The overall finding might suggest that the fish tended to react predominantly to the larger vortical structures and did not modulate their speed in response to small and short-lived turbulent structures. However, the distribution of t_{xx}/T_{xx} data hinted a bimodal distribution, with two dominant modes: $t_{xx}/T_{xx} \sim 0.12$ and $t_{xx}/T_{xx} \sim 2.8$, highlighted in Figure C.9B (horizontal dashed lines). The result might imply two preferential responses of fish to turbulence and turbulent structures. In one mode ($t_{xx}/T_{xx} < 1$), the fish would react passively to vortical structures, with their slow response possibly enabling them to be advected by the flow turbulence (e.g. in recirculation zones and secondary currents). In the second mode ($t_{xx}/T_{xx} > 1$), the fish would be proactive and respond very rapidly to a change in turbulent flow conditions. They would benefit from changes in instantaneous flow conditions to migrate upstream.

With carangiform propulsion, the fish propels itself largely with the caudal fin that is oscillated quasi-periodically. The fish tailbeat frequency data are reported in Figure C.9C in a dimensionless form $F \times k_s/V_x$ as a function of the dimensionless fish length L_f/k_s, where L_f is the total fish length and V_x is the time-averaged longitudinal velocity at the fish location. The data showed that the fish swimming tailbeat frequency spanned over a relatively narrow interval under the tested conditions (Eloy, 2012). Herein, the experimental data gave

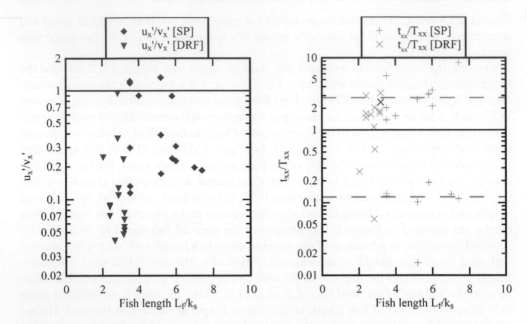

(A, Left) Ratio of fish speed to fluid velocity standard deviation u_x'/v_x' as a function of the dimensionless fish length L_f/k_s

(B, Right) Ratio of fish speed to fluid velocity auto-correlation time scales as a function of the dimensionless fish length L_f/k_s – dashed line indicates the two dominant modes

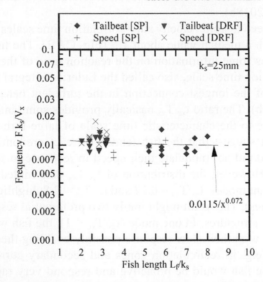

(C) Dimensionless frequencies of fish tailbeat and characteristic fish speed fluctuations $F \times k_s/V_x$ as functions of the dimensionless fish length L_f/k_s

Figure C.9 Dimensionless properties of swimming characteristics of juvenile silver perch (*Bidyanus bidyanus*) [SP] and adult Duboulay's rainbowfish (*Melanotaenia duboulayi*) [DRF] swimming along the rough sidewall in a channel with a rough bed and rough left sidewall for a relatively small flow: $Q = 0.0261$ m³/s – comparison with fluid velocity properties at the observation location

$F \times k_s/V_x \sim 0.01$ on average for both species. The present results showed some correlation in terms of the dimensionless fish length:

$$\frac{F \times k_s}{V_x} = 0.0115 \times \frac{L_f}{k_s} \qquad\qquad (C.6)$$

Equation (C.6) is compared to experimental data in Figure C.9C. The tailbeat frequency data were further compared to the characteristic frequency of the longitudinal fish speed. An example of the latter is illustrated in Figure C.8D. Figure C.8D shows a typical power spectrum density function of the longitudinal fish speed. All the frequency analyses presented a dominant characteristic frequency, as in Figure C.8D. The characteristic frequencies of the longitudinal fish speed are reported in Figure C.9C: the results showed that the characteristic fish speed frequency may be used as a proxy for the tailbeat frequency.

C.3.3   Discussion

Detailed full-scale laboratory data (Wang *et al.*, 2016a, 2016b; Cabonce *et al.*, 2017, 2018, 2019) highlighted a number of issues that deserve some discussion. The fish speed fluctuations were systematically smaller than the turbulent velocity fluctuations at the fish location (i.e. $u_x' < v_x'$). In turn, the fish accelerations were small and the corresponding inertial force was minimal. The fish swimming accelerations have some important implication in terms of energy expenditure required to swim against the current over a period of time (Plew *et al.*, 2007). Power is required to overcome friction and form drag (Videler, 1993), while additional energy is spent during acceleration phases. The combined power to overcome skin friction and form drag is proportional to the cube of fish speed relative to the mean fluid motion, that is, power $\propto (U_x + V_x)^3$, while the power required during acceleration periods is basically the fish mass time acceleration time relative to fish speed, that is, power $\propto m_f \times a_x \times (U_x + V_x)$ (Wang and Chanson, 2018a). A minimization of the fish accelerations would yield to smaller inertial forces and lesser energy consumption.

Further, a number of fish speed records suggested a secondary characteristic frequency, for example, about 16 Hz in Figure C.8D. While the primary frequency is likely to correspond to prolonged aerobic swimming, the secondary frequency might indicate some burst swimming or some short upstream motion generated by a few strong tailbeats. Visual observations showed a faster tailbeat frequency during sprint swimming. Further investigations could consider the characteristic fish acceleration frequencies, as well as the auto-correlation time scales of the fish acceleration and a quantitative description of the fish energy consumption during its upstream migration. In the view of ecologically friendly engineering design, which has initially motivated many studies, a comprehensive fish behavior study would be beneficial, including how fish sense fluid flow turbulence to select the optimum upstream path in turbulent flows. For example, many observations reported that small fish swim preferentially next to the bottom corners in rectangular channels – that is, in regions of low fluid velocity but very high turbulence and intense secondary currents.

Note

1 The very rough wall and invert treatment were installed along the whole flume length (Fig. C.4B). For the experiments with a rough sidewall and invert, the equivalent sand roughness height of the whole flume was $k_s \approx 20$ to 30 mm, comparable to the fish height h_f.

Computational fluid dynamics (CFD) modeling of a fish-friendly standard box culvert barrel

D.1 Presentation

To obtain a full picture of the detailed velocity field of the culvert flow through a barrel, hybrid modeling was conducted, consisting of physical experiments, one-dimensional (1D) theoretical calculation, and numerical computational fluid dynamics (CFD) calculation. In practice, a culvert structure can range from 1 m to 30 m in length, with a single cell being typically 0.5 m to 3 m in width and height. As a result, modeling large culvert cells using physical experiments in the laboratory is a challenge. Numerical CFD modeling was used as part of the current guideline development, coupled with 1D theoretical calculations (backwater calculations) to predetermine the free-surface level at the numerical inlet and outlet.

CFD modeling solves the Navier-Stokes equations of fluid motion using the numerical method. The Navier-Stokes equations, named after Claude-Louis Navier and George Gabriel Stokes, describe the motion of viscous fluid, coupled with the equations of conservation of mass. In its incompressible two-phase flow form, the equations can be written as:

$$\nabla \cdot \vec{u} = 0 \tag{D.1}$$

$$\rho\left(\frac{\partial \vec{u}}{\partial t} + (\vec{u} \cdot \nabla)\vec{u}\right) = \rho\vec{g} - \nabla p + \nabla \cdot \left[\mu\left(\nabla \vec{u} + \nabla^{\mathsf{T}}\vec{u}\right)\right] \tag{D.2}$$

where $\vec{u}$ is the velocity vector, p is the pressure, t is time, $\vec{g}$ is the gravity vector, ρ is the fluid density, and μ is the fluid viscosity.

Flow through a single box culvert barrel was simulated numerically over a wide range of flow rates, culvert sizes, aspect ratios, and tailwater levels. Table D.1 summarizes the flow conditions simulated numerically. Results of velocity field were analyzed, focusing particularly on local velocity and zones of low velocity. A relationship between the longitudinal velocity and the associated flow area was derived in dimensionless form. A number of numerical models were validated in detail, based upon physical measurements.

Computational fluid dynamics (CFD) modeling of a fish-friendly standard box culvert barrel

Table D.1 Numerical CFD modeling of culvert barrel flows: detailed flow conditions

Case	Design flow conditions									10% design flow conditions						Mesh grid number (longitudinal × vertical × transverse)
	Q_{des} (m³/s)	L (m)	S_o	d_{tw} (m)	afflux (m)	N_{cell}	B_{cell} (m)	D_{cell} (m)	Q_{cell} (m³/s)	Q (m³/s)	Q_{cell} (m³/s)	d_{tw} (m)	q (m²/s)	$(B/d)_{cell}$	V_{mean} (m/s)	
Gara river case	20	8	0	0.976	0.55	7	1.3	1	2.86	2	0.29	0.51	0.22	2.60	0.44	160×30×25
Exam paper case	4.8	14	0.0012	0.457	0.2	7	1	0.5	0.69	0.48	0.07	0.12	0.07	8.33	0.57	140×30×20
Laura river case	95	8	0.0015	2.195	0.45	10	2.4	2.4	9.50	9.5	0.95	0.51	0.40	4.71	0.80	160×48×48
Experiment 1	8		0				0.5	0.5			0.06	0.16	0.11	3.13	0.70	80×30×20
Experiment 2	12		0				0.5	0.5			0.03	0.10	0.05	5.00	0.52	240×30×25
Experiment 3	8		0.05				0.5	0.5			0.06	0.04	0.112	12.5	2.8	80×30×20
Experiment 4	8		0				0.5				0.11	0.30	0.22	1.67	0.75	80×30×20
Experiment 5	8		0				1	1			0.11	0.20	0.11	5.05	0.57	80×30×20
Experiment 6	19		0				0.7	0.5			0.10	0.10	0.14	4.24	0.87	380×30×20
Other 1	10	8	0.005		0.5	5	1	0.75	2.00	1	0.20	0.20	0.20	6.06	1.21	80×30×30

D.2   Methodology

CFD modeling of open channel flow through a culvert barrel was conducted with ANSYS (2017) Fluent version 18.0. A standard k-ε model was used to solve the flow turbulence (Rodi, 1995). For smooth turbulent flow through simplistic geometries, the flow physics is mostly dominated by boundary shear on the bottom boundary. A simplistic turbulence model such as a k-ε model is sufficient to resolve the velocity field with a relatively low computational cost. The simplified Reynolds Averaged Navier-Stokes equations are solved as:

$$\frac{\partial \rho}{\partial t} + \frac{\partial}{\partial x_j}\left(\rho u_j\right) = 0 \tag{D.3}$$

$$\frac{\partial \rho u_i}{\partial t} + \frac{\partial}{\partial x_j}\left(\rho u_i u_j\right) = -\frac{\partial p'}{\partial x_i} + \frac{\partial}{\partial x_j}\left[\mu_{eff}\left(\frac{\partial u_i}{\partial x_j} + \frac{\partial u_j}{\partial x_i}\right)\right] + S_M \tag{D.4}$$

where S_M is the sum of body forces, μ_{eff} is the effective viscosity representing flow turbulence, p' is the modified pressure, and the subscripts i and j represent properties in the i and j directions.

Based on the "eddy viscosity" concept proposed by Boussinesq (1897), the effective viscosity may be calculated as:

$$\mu_{eff} = \mu + \mu_t \tag{D.5}$$

where μ and μ_t are, respectively, the fluid viscosity and eddy (turbulent) viscosity.

The standard k-ε model used two transport equations to describe the turbulent viscosity. The two equations are for the turbulent kinetic energy k and dissipation ε, respectively (Launder and Spalding, 1974):

$$\frac{\partial}{\partial t}(\rho k) + \frac{\partial}{\partial x_i}(\rho k u_i) = \frac{\partial}{\partial x_j}\left[\left(\mu + \frac{\mu_t}{\sigma_k}\right)\frac{\partial k}{\partial x_j}\right] + G_k + G_b - \rho \varepsilon - Y_M + S_k \tag{D.6}$$

$$\frac{\partial}{\partial t}(\rho \varepsilon) + \frac{\partial}{\partial x_i}(\rho \varepsilon u_i) = \frac{\partial}{\partial x_j}\left[\left(\mu + \frac{\mu_t}{\sigma_\varepsilon}\right)\frac{\partial \varepsilon}{\partial x_j}\right] + C_{1\varepsilon}\frac{\varepsilon}{k}(G_k + C_{3\varepsilon}G_b) - C_{2\varepsilon}\rho\frac{\varepsilon^2}{k} + S_\varepsilon \tag{D.7}$$

where G_k represents the generation of turbulent kinetic energy due to the mean velocity gradient; G_b is the generation of turbulent kinetic energy due to buoyancy; Y_M represents the contribution of the fluctuating dilatation in compressible turbulence to the overall dissipation rate; $C_{1\varepsilon}$, $C_{2\varepsilon}$, and $C_{3\varepsilon}$ are constants; σ_k and σ_ε are the turbulent Prandtl numbers for k and ε, respectively; and S_k and S_ε are user-defined source terms.

The turbulent viscosity μ_t is computed by combining k and ε as:

$$\mu_t = \rho C_\mu \frac{k^2}{\varepsilon} \tag{D.8}$$

By default, ANSYS Fluent used the following values for constants: $C_{1\varepsilon} = 1.44$, $C_{2\varepsilon} = 1.92$, $C_\mu = 0.09$, $\sigma_k = 1.0$, and $\sigma_\varepsilon = 1.3$.

The two-phase flow interface in the culvert barrel was tracked by a volume of fluid (VOF) method (Hirt and Nichols, 1981). In VOF, a color function C was introduced, defined as 0 in one phase and 1 in the other. Herein, the primary phase was selected to be air (the lighter medium) and the secondary phase water. The function C is characterized by an advection equation:

$$\frac{\partial C}{\partial t} + \bar{u} \cdot \nabla C = 0 \tag{D.9}$$

Fluid properties such as density and viscosity are then calculated based on the respective fractions of the local color function.

The near-wall areas of the flow were treated by a built-in standard wall function in ANSYS Fluent. The wall function was based on the work of Launder and Spalding (1974) and is used widely in industrial flows. The log law was applied for near-wall regions to calculate the dimensionless velocity u* by:

$$u^* = \frac{1}{\kappa} \ln\left(Ey^*\right) \tag{D.10}$$

where:

$$u^* \equiv \frac{u_p C_\mu^{1/4} k_p^{1/2}}{\tau_w / \rho} \tag{D.11}$$

and:

$$y^* \equiv \frac{\rho C_\mu^{1/4} k_p^{1/2} y_P}{\mu} \tag{D.12}$$

κ is the von Karman constant and E is the empirical constant,[1] U_p is the mean velocity of the fluid at the wall-adjacent cell centroid P, k_p is the turbulence kinetic energy at the wall-adjacent cell centroid P, y_p is the distance from the centroid of the wall-adjacent cell to the wall P, and μ is the dynamic viscosity of the fluid.

The law of the wall (i.e. log law) for mean velocity is only valid for $30 < y^* < 300$ (Schlichting, 1979; Chanson, 2014). Herein, ANSYS Fluent employs the log law when $y^* > 11.225$. When the mesh yields $y^* < 11.225$ at the wall-adjacent cells, ANSYS Fluent applies the laminar stress-strain relationship:

$$u^* = y^* \tag{D.13}$$

For rough pipes and channels, the law of wall is modified to include a roughness effect as:

$$u^* = \frac{1}{\kappa} \ln\left(E y^*\right) - \Delta B \tag{D.14}$$

where ΔB is well correlated with the nondimensional roughness height K_s+ calculated as:

$$K_s+ = \frac{\rho k_s C_\mu^{1/4} k_p^{1/2}}{\mu} \tag{D.15}$$

with k_s the equivalent roughness height. In ANSYS Fluent, three distinct roughness regimes are employed. For a hydrodynamically smooth regime ($K_s+ \le 2.25$), $\Delta B = 0$. For a transitional regime ($2.25 \le K_s+ \le 90$):

$$\Delta B = \frac{1}{\kappa} \ln\left[\frac{K_s^+ - 2.25}{87.75} + C_s K_s^+\right] \times \sin\left\{0.4258\left(\ln K_s^+ - 0.811\right)\right\} \tag{D.16}$$

where C_s is a roughness constant (in this case $C_s = 0.5$ representing uniform roughness). In the fully rough regime ($K_s+ \geq 90$):

$$\Delta B = \frac{1}{\kappa} \ln\left[1 + C_s K_s^+\right] \tag{D.17}$$

The present study only focuses on smooth transitional turbulent flow due to the scope of the study being a box culvert with smooth concrete walls.

D.3   Numerical model configuration

The numerical domain representing a single box culvert barrel is illustrated in Figure D.1. Two barrel lengths were modeled (i.e. L_{barrel} = 8 m and 12 m). The width and height of the numerical domain were prescribed according to the internal width and height of the modeled

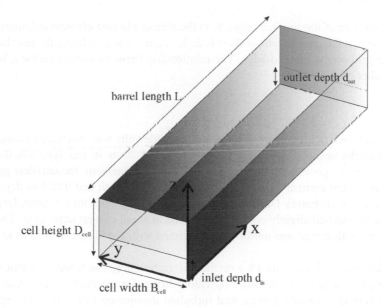

barrel length L

outlet depth d_{out}

cell height D_{cell}

X

y

cell width B_{cell}

inlet depth d_{in}

Figure D.1 3D sketch of numerical domain with color-coded boundaries; detailed boundary conditions listed in Table D.2

Table D.2 Color-coding of boundaries as shown in Figure D.1 and the boundary conditions

Color	Boundary name	Boundary condition	Remarks
Blue	Air inlet	Velocity inlet	Inlet velocity V_{in} = 0 m/s
Yellow	Water inlet	Velocity inlet	Inlet velocity V_{in} calculated from inlet discharge
Red	Walls	Wall	Roughness k_s = 0–0.001 m (i.e. smooth concrete) Uniform roughness
Green	Outlet (air and water)	Pressure outlet	Free-surface level at outlet d_{out} set from tailwater level d_{tw} ($d_{out} = d_{tw}$ in general)

culvert cells, B_{cell} and D_{cell}, respectively. The dimensions of the culvert cells are shown in Table D.1. The inlet plane, marked in yellow and blue in Figure D.1, was split into two velocity inlets: one for water (coded yellow) and one for air (coded blue). The outlet plane, coded green in Figure D.1, was a single outlet for both phases and set to be a pressure outlet. A free-surface level was required to set up the outlet for open channel flow, and this outlet depth d_{out} was prescribed according to the tailwater level for the modeled case. In general, $d_{out} \approx d_{tw}$, where d_{tw} was the tailwater depth in the floodplain downstream of the culvert barrel.

The numerical CFD modeling consisted of two stages: (a) transient flow simulation in a 3D culvert channel with coarse mesh; the coarse mesh consisted of uniform squares with 0.0 to −0.1 m grid size throughout the numerical domain and (b) transient flow simulation in a 3D culvert channel with refined mesh; the mesh was refined into nonuniform, gradually varied squares using a bias function:

$$\Delta = \sum_{0}^{i} \Delta_1 \times r^i \tag{D.18}$$

where Δ is the size of the meshed edge; Δ_1 is the size of the first element calculated using a bias factor bf; r is the growth rate; and i = 1, 2, 3, . . ., n − 1 with n being the number of divisions in the grid on the meshed edge. The relationship between growth factor r, bias factor bf, and number of division n is:

$$bf = r^{n-1} \tag{D.19}$$

A biased mesh with refinement near the walls and sidewalls was essential to simulate realistic flow patterns near the boundaries. A bias factor of 20 to 30 was typically used for all cases, resulting in a growth factor r = 1.1 to 1.2. After refinement, the smallest grid size in the transverse y and vertical z directions was between 0.001 m and 0.005 m depending on the size of the culvert barrel. Due to the computational cost and limit in time, large culvert structures were meshed slightly coarser[2] compared to small culvert structures. The mesh in the streamwise x direction was uniformly partitioned with a grid size of 0.05 m to 0.1 m for all cases.

All models were solved using a k-ε method for turbulence. The transient formulation was solved implicitly with a second-order upwind scheme for momentum, first-order upwind scheme for turbulent kinetic energy, and turbulent dissipation rate. The convergence was ensured by reducing residuals of all parameters to 10^{-4} or less. All simulations were run until the monitored free-surface level at a location stopped varying or only showed very small fluctuations and the conservation of mass was achieved between the inlet and outlet at the end of the transient simulation. Typically, the physical time taken to reach this stage was 60 to 90 s. The computation time for a complete run was approximately 12 to 24 hours on an HPC workstation (eight processors).

D.4   Numerical results and validation

A series of laboratory experiments was conducted by Cabonce *et al.* (2017, 2019) and Wang *et al.* (2016b, 2018) to model open channel flows through a box culvert barrel. Experiments were performed on a smooth bed (roughness height $k_s \approx 10^{-4}$ m), rough beds (roughness height $k_s \approx 0.02$–0.03 m), and triangular barrels installed on the bottom corner. Experimental measurements conducted on a smooth bed were used as a validation data set for the current numerical CFD models. Table D.3 presents the experimental flow conditions investigated by Cabonce et al. (2017, 2019) and the numerical model information corresponding to these flow cases. Herein, the inflow discharge Q is the flow through the experimental channel (i.e. a single culvert barrel); L is the length of the experimental channel/numerical domain; B is the internal width of the experimental channel; B_{cell} and D_{cell} are the internal width and height of the numerical domain, respectively; S_o is the channel/barrel slope for both experimental and numerical studies; d_1 and V_1 are, respectively, the depth and velocity measured at 8 m downstream of the experimental channel inletl V_{in} is the inlet velocity prescribed at the velocity inlet of the numerical model for the water phase; d_{out} is the free-surface level prescribed at the pressure outlet of the numerical model; Δx_{min}, Δy_{min}, and Δz_{min} are, respectively, the minimum mesh grid size in the longitudinal x, transverse y, and vertical z directions.

Figure D.2 shows the comparison between the numerically simulated free-surface elevation and the experimental measurements. The results demonstrated a good agreement

Table D.3 Comparison between experimental flow conditions (Cabonce, 2017, 2019) and numerical CFD models

	Q (m³/s)	L (m)	B (m)	S_o	d_1 (m)	V_1 (m/s)	Bed configuration
Cabonce et al. (2017, 2019)	0.0556	12	0.5	0	0.162	0.69	Smooth bed
	0.0261	12	0.5	0	0.096	0.54	Smooth bed

	Q (m³/s)	L (m)	B_{cell} (m)	D_{cell} (m)	S_o	V_{in} (m/s)	d_{out} (m)	Mesh grid density	Δx_{min} (m)	Δy_{min} (m)	Δz_{min} (m)
Present CFD study	0.0556	8	0.5	0.5	0	0.56	0.160	55,398 nodes 50,480 elements	0.100	0.001	0.002
	0.0261	12	0.5	0.5	0	0.50	0.096	212,992 nodes 197,625 elements	0.002	0.002	0.003

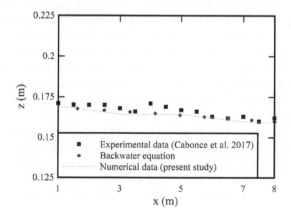

(A) Q = 0.056 m³/s

Figure D.2 Free-surface comparison between 1D numerical, CFD, and experimental results; experimental data from Cabonce et al. (2017) – flow condition: (A) Q = 0.056 m³/s, (B) Q = 0.026 m³/s

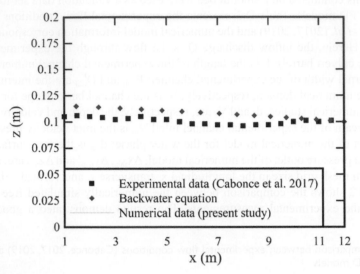

(B) Q = 0.026 m³/s

Figure D.2 (Continued)

between the 1D numerical, CFD, and experimental data in terms of free-surface elevation throughout the culvert channel. A key issue was to use a realistic tailwater depth d_{out}. The CFD model used a pressure outlet, which was very sensitive to the prescribed downstream free-surface level at the outlet. Herein, experimentally measured values were used at the outlet boundary to prescribe the tailwater depth, which was considered very important in reproducing the correct free-surface profile.

The vertical profiles of the longitudinal velocity component at different transverse locations were compared to experimental results for validation purposes. Typical outcomes are presented in Figures D.3 and D.4. Overall, the CFD data compared favorably to the physical results for all transverse locations, with the locations next to the sidewalls (0.08 m to wall) being better modeled than the center of the channel. The results showed an overall tendency of overestimating longitudinal velocity magnitudes by the CFD numerical model, especially toward the centerline of the channel. The maximum longitudinal velocity was overestimated by 10% using the CFD numerical model compared to the experimental data for the flow Q = 0.056 m³/s (Fig. D.4A).

Overall, the results showed the capacity of a CFD model to predict the 3D flow field in a smooth culvert barrel, which could be used to design a fish-friendly culvert. The systematic validation against physical data is critical to ascertain the performances of a numerical model and can be sensitive to a range of inflow conditions, boundary parameters, and grid mesh quality and size (Leng and Chanson, 2018; Zhang and Chanson, 2018).

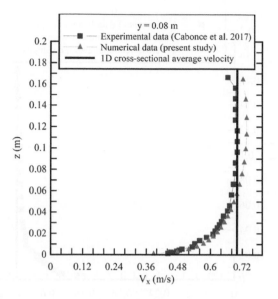

(A) Q = 0.056 m³/₃

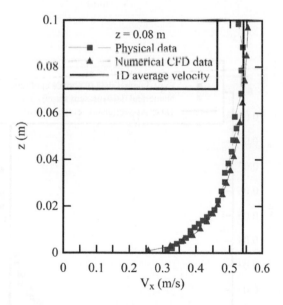

(B) Q = 0.026 m³/s

Figure D.3 Longitudinal velocity distribution in a box culvert barrel: comparison between 1D numerical, CFD numerical, and physical data; experiments by Cabonce *et al.* (2017); all measurements near the sidewall (0.08 m from wall) – flow conditions: (A) Q = 0.056 m³/s, (B) Q = 0.026 m³/s

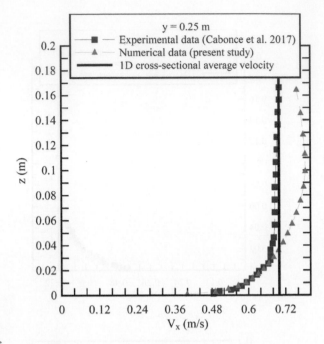

(A) Q = 0.056 m³/s

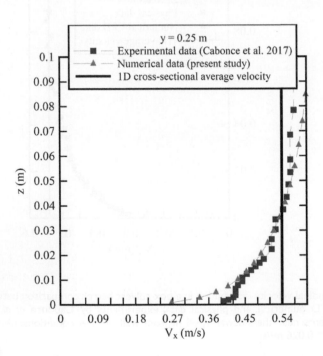

(B) Q = 0.026 m³/s

Figure D.4 Longitudinal velocity distribution in a box culvert barrel: comparison between 1D numerical, CFD numerical, and physical data; experiments by Cabonce *et al.* (2017); all measurements on the channel centerline (y = 0.25 m) – flow conditions: (A) Q = 0.056 m³/s, (B) Q = 0.026 m³/s

D.5   Area fraction of low-velocity zone: a numerical approach

Due to the large number of relevant design parameters (design discharge Q_{des}, tailwater level d_{tw}, maximum afflux, box cell configuration, etc.) and the case-specific nature of the culvert design (different targeted flood events for different regional councils), it is unrealistic to conduct CFD modeling for all possible design scenarios. Further, not all local governments and engineering companies have the capacity to conduct numerical CFD modeling. Hence, the calculation for the percentage of flow area of low-velocity zones must be generalized, with self-defined criteria for low velocity, independently of the hydrology requirement. That is, whether a targeted storm event is about 1:5 average recurrence interval (ARI) or 1:1 ARI. To achieve this, the present study examined the relationship between local velocity V_x and the associated flow area where the local velocity is less than that velocity. All data are compiled in a dimensionless form in Figure D.5. Details of flow conditions for data in Figure D.5 are presented in Table D.1.

Overall, all cases showed a similar trend, with quantitatively close results, albeit some scatter (Fig. D.5). The solid black curve represents the best-fit correlation of all data sets, whereas the two dashed lines illustrate the upper and lower bounds of the scatter. At an area fraction of 15%, the maximum difference between the two bounds of the data scatter was

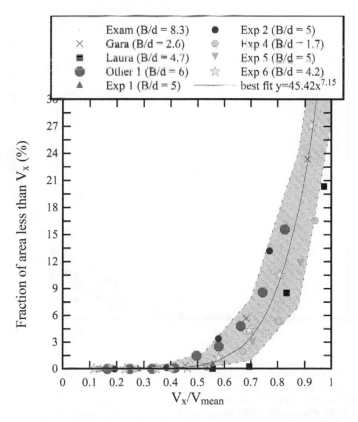

Figure D.5 Dimensionless area fraction of flow less than a relative longitudinal velocity V_x/V_{mean}, where V_{mean} is the bulk velocity (i.e. cross-sectional mean velocity in the barrel) – all cases compiled

approximately 10%. The quantitative differences between data sets seemed to show little relation to the aspect ratio B/d.

Figure D.6 compares present and past CFD and experimental works. The data are compared to an analytical solution for a two-dimensional turbulent flow, assuming a 1/N-th velocity distribution power law:

$$A = 100^{1-N} \times \left(\frac{N}{N+1}\right)^N \times \left(\frac{V_x}{V_{mean}}\right)^N \tag{D.20}$$

with A the percentage of flow area, V_{mean} the bulk velocity, and (V_x/V_{mean}) in percentage.

Equation D.20 is plotted for N = 4.5 in Figure D.6. The present CFD results showed a close agreement with past CFD data, although limited to only a few points. The experimental data

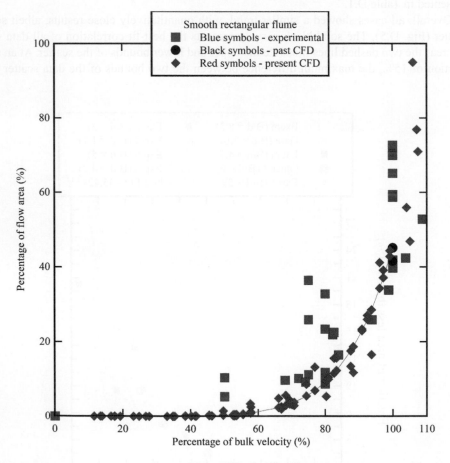

Figure D.6 Dimensionless area fraction of flow less than a relative longitudinal velocity V_x/V_{mean}, where V_{mean} is the bulk velocity (i.e. cross-sectional mean velocity in the barrel) – all cases compared to past CFD (Naot and Rodi, 1982), experimental studies (Cabonce et al., 2017; Xie, 1998; Macintosh, 1990; Nezu and Rodi, 1985; Nikuradse, 1926), and Equation (D.20), assuming N = 4.5

showed overall a larger area fraction for the same relative velocity compared to CFD data. The lower bound of experimental data scatter agreed closely with the upper bound of CFD data scatter.

It is worth noting a few advantages of using such a dimensionless plot (Fig. D.6). First, the plot is independent of hydrological implication, which could vary based on the requirements of different councils and sites. Second, the results are independent of the barrel culvert cell size and downstream tailwater conditions.

Notes

1 In ANSYS Fluent, $\kappa = 0.4187$ and $E = 9.793$.
2 That is, the grid size of 0.05 m in all three directions.

Appendix E

On alternatives to improve the upstream passage of small-body-mass fish, including retrofitting

E.1 Presentation

During the last decades, concerns about the ecological impact of culverts on stream connectivity have led to some evolution in design (Chorda *et al.*, 1995; Warren and Pardew, 1998; Hotchkiss and Frei, 2007). The impact in terms of fish passage may adversely affect the upstream and downstream ecosystems (Briggs and Galarowicz, 2013). Common culvert fish passage barriers encompass perched outlets with an excessive vertical drop at the culvert outlet, high velocities and turbulence in the barrel, debris accumulation at the culvert inlet, and standing waves in the inlet and outlet (Behlke *et al.*, 1991; Olsen and Tullis, 2013; Wang *et al.*, 2018). In this appendix, a review of different wall boundary treatments and appurtenances to improve the upstream passage of small fish in a culvert barrel is presented, based upon recent detailed hydrodynamic measurements in near-full-scale culvert barrel flumes (Fig. E.1 and Table E.1). Importantly, these boundary treatments were selected to have a relatively small impact on the discharge capacity of the culvert at design discharge.

All the experimental works were conducted in rectangular channels with discharges typical of less-than-design discharges, and fish endurance tests were conducted for a limited range of discharges and configurations (Appendix C). The data were complemented by detailed numerical computational fluid dynamics (CFD) modeling (Zhang and Chanson, 2018; Leng and Chanson, 2018). Noteworthy, the comparison between different boundary treatments is developed with identical water discharge in line with engineering design practices. Further, Appendix F presents several movies of fish swimming in the bottom corners with various boundary treatments: the movies CIMG1655.mov, CIMG2672.mov, CIMG1647.mov, CIMG2725.mov, and CIMG1651.mov are in a smooth box culvert channel; the movies CIMG1497.mov and CIMG2523.mov are in a channel with a very rough invert; and the movies CIMG1243.mov, CIMG1268.mov, CIMG1409.mov, and CIMG1410.mov are in a box culvert barrel with large asymmetrical roughness.

E.2 Basic results

Small-body-mass fish swim primarily next to the culvert barrel corners and sidewalls (Appendix C), although negative-wake flows may disorientate small fish (Cabonce *et al.*, 2018, 2019; Duguay *et al.*, 2018). Low-positive-velocity zones (LPVZs) suitable to small-bodied fish passage must fulfill:

$$0 < V_x < U_{fish} \tag{E.1}$$

where V_x is the local time-averaged longitudinal velocity component and U_{fish} is a characteristic fish speed (Chanson and Leng, 2018).

Fish navigability in a culvert barrel depends not only on the size of the LPVZ but also upon the connectivity between low-velocity zones (LVZs). In plain terms, long contiguous reaches of LPVZs that meet certain velocity criteria (e.g. $0 < V_x < U_{fish}$) are naturally more traversable than multiple, separate patches of LVZs (Zhang and Chanson, 2018). This illustrated in Figure E.2, in which a closer relative baffle-to-baffle spacing reduces the longitudinal velocity variation and improves the connectivity between LVZs, with benefits in terms of fish traversability.

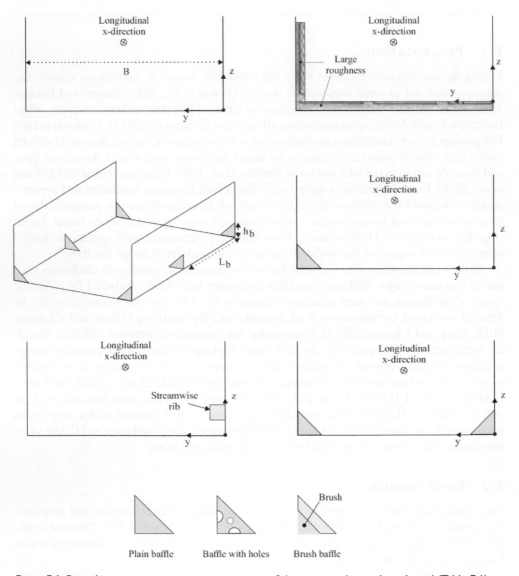

Figure E.1 Boundary treatment to improve upstream fish passage in box culvert barrel (Table E.1)

Table E.1 Hydrodynamic investigations of the box culvert barrel boundary treatment to assist upstream fish passage

Boundary treatment	Boundary conditions	Q (m³/s)	L (m)	B (m)	Fish species	Reference
Smooth channel	Smooth PVC bed and glass sidewalls	0.0261 0.0556	12	0.5	Duboulay's rainbowfish (Melanotaenia duboulayi), juvenile silver perch (Bidyanus bidyanus)	Wang et al. (2018) and Cabonce et al. (2019)
Rough-bed channel	Rough bed, smooth sidewalls	0.0261	12	0.5	Duboulay's rainbowfish (Melanotaenia duboulayi)	Wang et al. (2016a, 2018)
Asymmetrically roughened channel	Rough left wall, rough bed, smooth right wall	0.0261	12	0.4785	Duboulay's rainbowfish (Melanotaenia duboulayi), juvenile silver perch (Bidyanus bidyanus)	Wang et al. (2016a, 2018)
Small triangular corner baffles	Triangular corner baffles along left wall only, h_b = 0.067 and 0.133 m, L_b = 0.67 and 1.33 m	0.0261 0.0556	12	0.5	Juvenile silver perch (Bidyanus bidyanus)	Cabonce et al. (2018, 2019)
	Triangular corner baffles along both walls, h_b = 0.067 and 0.133 m, L_b = 0.67 and 1.33 m	0.0556			—	Cabonce et al. (2018) and Freire et al. (2018)
Ventilated triangular baffles (both sides)	Brush baffles along both walls, h_b = 0.133 m, L_b = 1.33 m	0.0556	12	0.50	—	Freire et al. (2018)
	Baffles with three holes along both walls, h_b = 0.133 m, L_b = 1.33 m				—	Freire et al. (2018)
Asymmetrical ribbed channel	0.050m × 0.050 m longitudinal rib placed along right wall 0.050 m above bed	0.0261 0.0556 0.100	15	0.50	—	Sanchez et al. (2018)

Notes: B: channel width; h_b: baffle height; L: channel length; L_b: longitudinal baffle spacing; Q: water discharge; horizontal channels: S_o = 0; light shade: fish endurance testing (Appendix C).

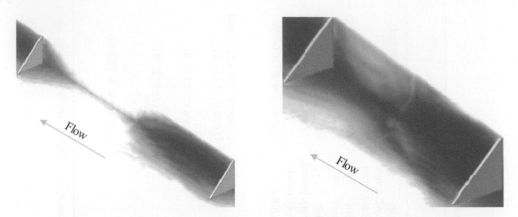

Figure E.2 Low-velocity zones in box culvert barrel with small triangular corner baffles on one side –
(Left) Q = 0.0556 m³/s, B_{cell} = 0.50 m, h_b = 0.067 m, L_b = 0.67 m; (Right) Q = 0.0556 m³/s,
B_{cell} = 0.50 m, h_b = 0.133 m, L_b = 0.67 m – LVZ (darker shades) defined as U/U_{mean} < 0.5
(after Zhang and Chanson, 2018)

The performances of several types of boundary treatments may be assessed in terms of the size of the LPVZ: $0 < V_x < 0.5 \times V_{mean}$ and their longitudinal distribution. Figure E.3 presents a comparison based upon detailed hydrodynamic data obtained in near-full-scale culvert barrel flumes for the same discharge.[1] In Figure E.3, the data are presented for three successive baffle spacings. The results show marked differences between the different types of boundary treatments (Fig. E.3). In the smooth and rough flumes, the flow resistance is regularly distributed and flow separation is negligible. Continuous LPVZs are provided next to the channel boundaries and in the corner regions, with a large LPVZ area fraction in the asymmetrical roughened channel. With small triangular corner baffles, flow separation occurs at each baffle edge, followed by a negative-velocity zone (NVZ), that is, a recirculation region. Immediately downstream of the small baffles, the size of LPVZ may be comparable to that in the rough wall and bed flume, but only for a short distance. A lack of longitudinal LPVZ interconnection is clearly evidenced in some cases, in Figure E.3, and consistent with the flow visualization in Figure E.2 (left). In contrast, the ribbed channel provides a smaller LPVZ, and a key difference is the well-marked, highly turbulent LVZ beneath the sidewall rib. The application of longitudinal sidewall rib treatment must be considered with the utmost care. A number of practical considerations yield major technical challenges and, in many instances, alternative designs should be preferred (Sanchez *et al.*, 2018).

In summary, a comparison was conducted between various boundary treatments selected to have a minimal impact on the discharge capacity of the culvert at large discharges. The comparative analysis of detailed hydrodynamic measurements with different boundary treatments suggests that the requirements for continuous, sizable LPVZs (e.g. $0 < V_x < 0.5 \times V_{mean}$) suitable to small-bodied fish might be best fulfilled with asymmetrically placed large roughness[2] culvert barrels sketched in Figure E.1 (top right) and shown in Figure E.4.

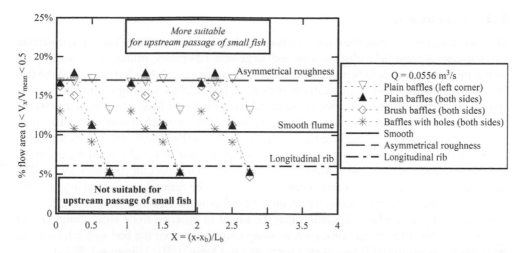

Figure E.3 Comparison in terms of longitudinal variation of fractions of LPVZ (i.e. $0 < V_x < 0.5 \times V_{mean}$) between different boundary treatments in a 12 m long and 0.5 m wide culvert barrel channel – experimental flow conditions listed in Table E.1, all data obtained in the fully developed flow region for the same discharge $Q = 0.0556$ m³/s, data shown over three longitudinal baffle spacings ($3 \times L_b$)

Figure E.4 Juvenile silver perch (*Bidyanus bidyanus*) swimming upstream in a culvert barrel with asymmetrical large roughness boundaries (right smooth glass sidewall, left rough wall, and rough invert) – flow conditions: $Q = 0.261$ m³/s, $V_{mean} = 0.43$ m/s, $S_o = 0$, flow direction from right to left

E.3   Discussion

With weak-swimming fish, the excessive barrel velocities are too often a barrier to upstream fish passage. Recent field observations and large-size laboratory studies documented fish swimming and behavior in box culvert barrels. All the biological data indicated that the fish swim preferentially close to sidewalls in regions of low velocity and high turbulence intensity (Appendix C). A key flow zone is the barrel corner region where secondary currents are strong. Next to a sidewall, the channel flow is retarded, and the result is some secondary flow motion generated at a right angle to the longitudinal current (Bradshaw, 1987; Sanchez et al., 2018). In the channel corner, a transverse flow is initiated and directed toward the corner as a direct result of turbulent shear stress gradients normal to the bisector (Gessner, 1973). The interactions between the transverse shear gradient along the corner bisector and longitudinal flow motion induce energy losses, which must be compensated for by some transverse flow. More generally, secondary circulation may be found in cross-sections with abrupt spatial variations in boundary conditions, such as sharp corners between the bed and sidewall and between the sidewall and free-surface (Tominaga and Nezu, 1991; Uijttewaal, 2014).

In the presence of different types of boundary treatments, the observations showed the preferred "spots" that the fish exploit, namely regions of low velocity and high turbulence with intense secondary motion. Irrespective of the boundary treatment, detailed observations indicated that fish spent two-thirds of their time in the bottom corners and nearly 90% of their time next to the sidewalls and bottom corners altogether (Appendix C). Present knowledge suggests further that the invert and sidewall roughness may be scaled to the fish dimensions, with some fish best interacting with large vortices when the ratio of eddy size to fish length is close to unity.

E.4   Concluding remarks: and what about turbulence?

Although leading scholars have discussed the "mechanics of fish–turbulence interactions" (Nikora et al., 2003, p. 1380) or "the role of turbulence on biotic communities" (Maddock et al., 2013, p. 433), what do we really know about turbulence?

Nobel Laureate Richard P. Feynman reminded us that "turbulence is the most important unsolved problem of classical physics" (Feynman et al., 1964), and Professor Peter Bradshaw added that "turbulence and its measurement are both controversial subjects" (Bradshaw, 1971, p. xii). Researchers cannot be complacent about turbulence because it is ubiquitous in nature: "turbulence is the most common, the most important and most complicated kind of flow motion" (Bradshaw, 1971, p. xi) and "many of its seemingly simple questions remain unanswered" (Smits and Marusic, 2013, p. 25). In a wall-bounded channel, such as a culvert barrel, turbulence is produced by friction at the solid boundaries, and the turbulent flow is constrained by the no-slip condition[3] at the barrel walls and invert. The flow region where the hydrodynamic properties are affected by boundary friction is broadly called the "boundary layer" (Schlichting, 1979; Bailly and Comte-Bellot, 2015), a concept first introduced by Ludwig Prandtl (1904).

A boundary layer is defined as the region where the flow properties are affected by boundary friction. In the fully developed flow region of a culvert barrel channel, the "boundary layer" region occupies the whole flow area, as shown experimentally (Appendix C and Table E.1). Physically, however, there were fundamental differences in the key turbulent processes between different boundary treatments that cannot be ignored. Let us consider a number of boundary treatments listed in Table E.1. With a smooth culvert barrel channel, the dominant mechanism of energy dissipation is the boundary skin friction, with a small secondary current of Prandtl's second kind in the bottom corners (Einstein and Li, 1956). With small triangular corner baffles, the flow field is dominated by fluid streamline separation at the edge of each baffle (Cabonce et al., 2019), with a negative wake behind and boil of the first kind (Schlichting, 1979). In presence of a longitudinal rib, strong secondary circulation of Prandtl's second kind occurs, linked to the development of large streamwise vortices and surface longitudinal streaks (Levi, 1965; Sanchez et al., 2018).

The interpretation of the turbulence typology is critical to a successful boundary treatment conducive to upstream passage of weak-swimming fish in a culvert barrel. A precise knowledge of the entire three-dimensional velocity field is essential, because the rate of work and energy required by a fish to thrust itself against the water discharge is proportional to the cube of the local fluid velocity, that is, V_x^3 (Wang and Chanson, 2018a). In plain terms, an in-depth understanding of the turbulent flow field constitutes a core requirement to comprehend the fish–fluid interactions and is a prerequisite for physically based mitigation measures of the ecological impact of culverts in terms of upstream fish passage.

Notes

1 All the data were measured in fully developed flow regions.
2 That is, with an equivalent roughness height comparable to the fish height: $k_s \sim h_f$, as tested by Wang et al. (2016a).
3 That is, $V_x = 0$ at a fixed rigid boundary.

Digital appendix: video movies of culvert operation and small-bodied fish swimming in box culvert channels

F.1 Presentation

The book aims to facilitate the advancement and diffusion of knowledge of the flow physics in box culverts and fish kinematics in culvert barrels. It contains a number of photographs and movies showing a range of flow phenomena and interactions with fish. This appendix presents a series of movies of box culvert operation and high-speed video movies of fish swimming taken by the author's research group. The movies illustrate some aspects of fish swimming in a full-scale box culvert barrel. Each movie is described below. Table F.1 summarizes the hydraulic engineering conditions of the high-speed video movies of small-bodied fish swimming in box culvert channels.

All the movies are copyright Hubert Chanson.

Video movie access

The video movies are accessible through the publisher's website at: www.routledge.com/cw/chanson

F.2 Video movies of standard box culvert operation

Title:	Box culvert operation at full capacity
Filename:	CIMG5048.mov
Duration:	10 s
Camera details:	Casio Exilim EX-10, movie mode set at 30 fps, full HD resolution 1920 × 1080 pixels
Description:	Box culvert operation at full capacity with inlet control and submerged inlet. Laboratory model at the University of Queensland. Flow conditions: $Q = 0.012$ m³/s ($\Delta h = 0.109$ m), $B_{cell} = 0.150$ m, $D_{cell} = 0.107$ m, $d_{hw} = 0.146$ m, $d_{hw} = 0.065$ m, submerged inlet.
Relevant book sections:	Chapter 2
Reference:	Chanson, H. (2004). "The Hydraulics of Open Channel Flow: An Introduction." *Butterworth-Heinemann*, 2nd edition, Oxford, UK, 630 pages.
Title:	Box culvert operation at 20% design discharge

Filename:	CIMG6255.jpg
Duration:	18 s
Camera details:	Casio Exilim EX-10, movie mode set at 30 fps, full HD resolution 1920×1080 pixels
Description:	Box culvert operation at 20% design discharge with outlet control and free-surface inlet. Laboratory model at the University of Queensland.
	Flow conditions: Q = 0.002 m³/s (Δh = 0.015 m), B_{cell} = 0.150 m, D_{cell} = 0.107 m, d_{hw} = 0.045 m, d_{hw} = 0.030 m, submerged inlet.
Relevant book sections:	Chapter 2
Reference:	Chanson, H. (2004). "The Hydraulics of Open Channel Flow: An Introduction." *Butterworth-Heinemann*, 2nd edition, Oxford, UK, 630 pages.

F.3   High-speed video movies of small-bodied fish swimming in box culvert channels

Fish swimming observations were conducted in a 12 m long and 0.5 m wide box culvert barrel channel (Table F.1). All experiments were conducted based upon protocols developed with biologists in accordance with strict animal ethical standards (Wang *et al.*, 2016a; Cabonce *et al.*, 2017, 2018, 2019).

F.3.1   Smooth culvert barrel channel

Title:	Duboulay's rainbowfish swimming upstream in a smooth box culvert barrel flume
Filename:	CIMG1655.mov
Duration:	64 s
Camera details:	Casio Exilim EX-10, movie mode set at 240 fps (replay 30 fps), 512×384 pixels
Description:	Duboulay's rainbowfish (*Melanotaenia duboulayi*) swimming upstream in the left bottom corner of a smooth box culvert barrel flume (12 m long, 0.5 m wide) for q = 0.052 m²/s, V_{mean} = 0.54 m/s, θ = 0 – fish characteristics: L_f = 69 mm, m_f = 2.7 g
Relevant book sections:	Chapter 4
Reference:	Wang, H., Chanson, H., Kern P., and Franklin, C. (2016). "Culvert Hydrodynamics to enhance Upstream Fish Passage: Fish Response to Turbulence." *Proceedings of 20th Australasian Fluid Mechanics Conference*, Australasian Fluid Mechanics Society, G. Ivey, T. Zhou, N. Jones, S. Draper Editors, Perth WA, Australia, 5–8 December, Paper 682, 4 pages.
	Wang, H., and Chanson, H. (2018). "On Upstream Fish Passage in Standard Box Culverts: Interactions between Fish and Turbulence." *Journal of Ecohydraulics*, IAHR, Vol. 3, No. 1, pp. 18–29 (DOI: 10.1080/24705357.2018.1440183).

Title:	Juvenile silver perch swimming upstream in a smooth box culvert barrel flume
Filename:	CIMG2672.mov
Duration:	8 min 21 s

Camera details:	Casio Exilim EX-10, movie mode set at 240 fps (replay 30 fps), 512×384 pixels
Description:	Juvenile silver perch (*Bidyanus bidyanus*) swimming upstream next to the left bottom corner of a smooth culvert barrel flume for $q = 0.112$ m²/s, $V_{mean} = 0.69$ m/s, $\theta = 0$ – fish characteristics: $L_f = 55$ mm, $m_f = 1.5$ g
Relevant book sections:	Chapter 4
Reference:	Wang, H., Chanson, H., Kern, P., and Franklin, C. (2016). "Culvert Hydrodynamics to enhance Upstream Fish Passage: Fish Response to Turbulence." *Proceedings of 20th Australasian Fluid Mechanics Conference*, Australasian Fluid Mechanics Society, G. Ivey, T. Zhou, N. Jones, S. Draper Editors, Perth WA, Australia, 5–8 December, Paper 682, 4 pages.
	Wang, H., and Chanson, H. (2018). "On Upstream Fish Passage in Standard Box Culverts: Interactions between Fish and Turbulence." *Journal of Ecohydraulics*, IAHR, Vol. 3, No. 1, pp. 18–29 (DOI: 10.1080/24705357.2018.1440183).

Title:	Duboulay's rainbowfish swimming upstream in a smooth box culvert barrel flume
Filename:	CIMG1647.mov
Duration:	2 min 53 s
Camera details:	Casio Exilim EX-10, movie mode set at 240 fps (replay 30 fps), 512×384 pixels
Description:	Duboulay's rainbowfish (*Melanotaenia duboulayi*) swimming upstream in the right bottom corner of a smooth box culvert barrel flume (12 m long, 0.5 m wide) for $q = 0.052$ m²/s, $V_{mean} = 0.54$ m/s, $\theta = 0$ – fish characteristics: $L_f = 37$ mm, $m_f = 0.7$ g
Relevant book sections:	Chapter 4
Reference:	Wang, H., Chanson, H., Kern, P., and Franklin, C. (2016). "Culvert Hydrodynamics to enhance Upstream Fish Passage: Fish Response to Turbulence." *Proceedings of 20th Australasian Fluid Mechanics Conference*, Australasian Fluid Mechanics Society, G. IVEY, T. Zhou, N. Jones, S. Draper Editors, Perth WA, Australia, 5–8 December, Paper 682, 4 pages.
	Wang, H., and Chanson, H. (2018). "On Upstream Fish Passage in Standard Box Culverts: Interactions between Fish and Turbulence." *Journal of Ecohydraulics*, IAHR, Vol. 3, No. 1, pp. 18–29 (DOI: 10.1080/24705357.2018.1440183).

Title:	Juvenile silver perch swimming upstream in a smooth box culvert barrel flume
Filename:	CIMG2725.mov
Duration:	9 min 37 s
Camera details:	Casio Exilim EX-10, movie mode set at 240 fps (replay 30 fps), 512×384 pixels
Description:	Juvenile silver perch (*Bidyanus bidyanus*) swimming upstream next to the left bottom corner of a smooth culvert barrel flume for $q = 0.112$ m²/s, $V_{mean} = 0.69$ m/s, $\theta = 0$ – fish characteristics: $L_f = 91$ mm, $m_f = 2.5$ g – almost stationary swimming in bottom corner of channel

Relevant book sections:	Chapter 4
Reference:	Wang, H., Chanson, H., Kern, P., and Franklin, C. (2016). "Culvert Hydrodynamics to enhance Upstream Fish Passage: Fish Response to Turbulence." *Proceedings of 20th Australasian Fluid Mechanics Conference*, Australasian Fluid Mechanics Society, G. IVEY, T. Zhou, N. Jones, S. Draper Editors, Perth WA, Australia, 5–8 December, Paper 682, 4 pages.
	Wang, H., and Chanson, H. (2018). "On Upstream Fish Passage in Standard Box Culverts: Interactions between Fish and Turbulence." *Journal of Ecohydraulics*, IAHR, Vol. 3, No. 1, pp. 18–29 (DOI: 10.1080/24705357.2018.1440183).

Title:	Duboulay's rainbowfish swimming upstream in a smooth box culvert barrel flume
Filename:	CIMG1651.mov
Duration:	2 min 13 s
Camera details:	Casio Exilim EX-10, movie mode set at 240 fps (replay 30 fps), 512 × 384 pixels
Description:	Duboulay's rainbowfish (*Melanotaenia duboulayi*) swimming upstream in the right bottom corner of a smooth box culvert barrel flume (12 m long, 0.5 m wide) for q = 0.052 m²/s, V_{mean} = 0.54 m/s, θ = 0 – fish characteristics: L_f = 61 mm, m_f = 2.2 g
Relevant book sections:	Chapter 4
Reference:	Wang, H., Chanson, H., Kern, P., and Franklin, C. (2016). "Culvert Hydrodynamics to enhance Upstream Fish Passage: Fish Response to Turbulence." *Proceedings of 20th Australasian Fluid Mechanics Conference*, Australasian Fluid Mechanics Society, G. Ivey, T. Zhou, N. Jones, S. Draper Editors, Perth WA, Australia, 5–8 December, Paper 682, 4 pages.
	Wang, H., and Chanson, H. (2018). "On Upstream Fish Passage in Standard Box Culverts: Interactions between Fish and Turbulence." *Journal of Ecohydraulics*, IAHR, Vol. 3, No. 1, pp. 18–29 (DOI: 10.1080/24705357.2018.1440183).

F.3.2 Culvert barrel channel with very rough bed and smooth sidewalls

Title:	Duboulay's rainbowfish swimming upstream in a box culvert barrel flume with large bed roughness
Filename:	CIMG1497.mov
Duration:	13 min 23 s
Camera details:	Casio Exilim EX-10, movie mode set at 240 fps (replay 30 fps), 512 × 384 pixels
Description:	Duboulay's rainbowfish (*Melanotaenia duboulayi*) swimming upstream in the right bottom corner of a very rough bed culvert barrel flume (12 m long, 0.5 m wide) for q = 0.052 m²/s, V_{mean} = 0.54 m/s, θ = 0 – fish characteristics: L_f = 71 mm, m_f = 2.7 g – almost stationary swimming in bottom corner of channel, with the fish using the large rugosity of the bed to shelter
Relevant book sections:	Chapter 4

Reference:	Wang, H., Chanson, H., Kern, P., and Franklin, C. (2016). "Culvert Hydrodynamics to enhance Upstream Fish Passage: Fish Response to Turbulence." *Proceedings of 20th Australasian Fluid Mechanics Conference*, Australasian Fluid Mechanics Society, G. Ivey, T. Zhou, N. Jones, S. Draper Editors, Perth WA, Australia, 5–8 December, Paper 682, 4 pages.
	Wang, H., and Chanson, H. (2018). "On Upstream Fish Passage in Standard Box Culverts: Interactions between Fish and Turbulence." *Journal of Ecohydraulics*, IAHR, Vol. 3, No. 1, pp. 18–29 (DOI: 10.1080/24705357.2018.1440183).

Title:	Duboulay's rainbowfish swimming upstream in a box culvert barrel flume with large bed roughness
Filename:	CIMG1523.mov
Duration:	13 min 57 s
Camera details:	Casio Exilim EX-10, movie mode set at 240 fps (replay 30 fps), 512 × 384 pixels
Description:	Duboulay's rainbowfish (*Melanotaenia duboulayi*) swimming upstream in the right bottom corner of a very rough bed culvert barrel flume (12 m long, 0.5 m wide) for $q = 0.052$ m²/s, $V_{mean} = 0.54$ m/s, $\theta = 0$ – fish characteristics: $L_f = 78$ mm, $m_f = 3.3$ g – almost stationary swimming in bottom corner of channel, with the fish using the large rugosity of the bed to shelter
Relevant book sections:	Chapter 4
Reference:	Wang, H., Chanson, H., Kern, P., and Franklin, C. (2016). "Culvert Hydrodynamics to enhance Upstream Fish Passage: Fish Response to Turbulence." *Proceedings of 20th Australasian Fluid Mechanics Conference*, Australasian Fluid Mechanics Society, G. Ivey, T. Zhou, N. Jones, S. Draper Editors, Perth WA, Australia, 5–8 December, Paper 682, 4 pages.
	Wang, H., and Chanson, H. (2018). "On Upstream Fish Passage in Standard Box Culverts: Interactions between Fish and Turbulence." *Journal of Ecohydraulics*, IAHR, Vol. 3, No. 1, pp. 18–29 (DOI: 10.1080/24705357.2018.1440183).

F.3.3 Asymmetrical large roughness culvert barrel channel

Title:	Juvenile silver perch swimming upstream in a box culvert barrel flume with asymmetrical large roughness
Filename:	CIMG1243.mov
Duration:	18 s
Camera details:	Casio Exilim EX-10, movie mode set at 240 fps (replay 30 fps), 512 × 384 pixels
Description:	Juvenile silver perch (*Bidyanus bidyanus*) swimming upstream in the right bottom corner of a culvert barrel flume (12 m long, 0.5 m wide) equipped with asymmetrical large roughness for $q = 0.052$ m²/s, $V_{mean} = 0.54$ m/s, $\theta = 0$ – fish characteristics: $L_f = 200$ mm, $m_f = 125.5$ g – the fish swims next to the right smooth sidewall. This fish was large enough to be comfortable swimming in all parts of the flume, although it swam mainly in the bottom floor region. After it discovered the (left) rough corner, the fish remained there for the rest of filming. Note the free-surface trough in the near wake of the fish body during active swimming

Relevant book sections:	Chapter 4
Reference:	Wang, H., Chanson, H., Kern, P., and Franklin, C. (2016). "Culvert Hydrodynamics to enhance Upstream Fish Passage: Fish Response to Turbulence." *Proceedings of 20th Australasian Fluid Mechanics Conference*, Australasian Fluid Mechanics Society, G. Ivey, T. Zhou, N. Jones, S. Draper Editors, Perth WA, Australia, 5–8 December, Paper 682, 4 pages. Wang, H., Uys, W., and Chanson, H. (2018). "Alternative Mitigation Measures for Fish Passage in Standard Box Culverts: Physical Modelling." *Journal of Hydro-environment Research*, IAHR, Vol. 19, pp. 214–223 (DOI: 10.1016/j.jher.2017.03.001).

Title:	Juvenile silver perch swimming upstream in a box culvert barrel flume with asymmetrical large roughness
Filename:	CIMG1268.mov
Duration:	3 min 14 s
Camera details:	Casio Exilim EX-10, movie mode set at 240 fps (replay 30 fps), 512 × 384 pixels
Description:	Juvenile silver perch (*Bidyanus bidyanus*) swimming upstream in the right bottom corner of a culvert barrel flume (12 m long, 0.5 m wide) equipped with asymmetrical large roughness for q = 0.052 m²/s, V_{mean} = 0.54 m/s, θ = 0 – fish characteristics: L_f = 130 mm, m_f = 29.1 g – the fish swims next to the left rough sidewall. The fish swam for a significant length of time in the corner of the flume. It gradually made its way up the flume, sometimes hiding in the roughness holes
Relevant book sections:	Chapter 4
Reference:	Wang, H., Chanson, H., Kern, P., and Franklin, C. (2016). "Culvert Hydrodynamics to enhance Upstream Fish Passage: Fish Response to Turbulence." *Proceedings of 20th Australasian Fluid Mechanics Conference*, Australasian Fluid Mechanics Society, G. Ivey, T. Zhou, N. Jones, S. Draper Editors, Perth WA, Australia, 5–8 December, Paper 682, 4 pages. Wang, H., Uys, W., and Chanson, H. (2018). "Alternative Mitigation Measures for Fish Passage in Standard Box Culverts: Physical Modelling." *Journal of Hydro-environment Research*, IAHR, Vol. 19, pp. 214–223 (DOI: 10.1016/j.jher.2017.03.001).

Title:	Duboulay's rainbowfish swimming upstream in a box culvert barrel flume with asymmetrical large roughness
Filename:	CIMG1409.mov
Duration:	5 min 19 s
Camera details:	Casio Exilim EX-10, movie mode set at 240 fps (replay 30 fps), 512 × 384 pixels
Description:	Duboulay's rainbowfish (*Melanotaenia duboulayi*) swimming upstream in the right bottom corner of a culvert barrel flume (12 m long, 0.5 m wide) equipped with asymmetrical large roughness for q = 0.052 m²/s, V_{mean} = 0.54 m/s, θ = 0 – fish characteristics: L_f = 80 mm, m_f = 4.4.1 g – the fish swims next to the left rough sidewall, in an almost stationary manner, using the corner roughness arrangement to shelter

Relevant book sections:	Chapter 4
Reference:	Wang, H., Chanson, H., Kern, P., and Franklin, C. (2016). "Culvert Hydrodynamics to enhance Upstream Fish Passage: Fish Response to Turbulence." *Proceedings of 20th Australasian Fluid Mechanics Conference*, Australasian Fluid Mechanics Society, G. Ivey, T. Zhou, N. Jones, S. Draper Editors, Perth WA, Australia, 5–8 December, Paper 682, 4 pages.
	Wang, H., Uys, W., and Chanson, H. (2018). "Alternative Mitigation Measures for Fish Passage in Standard Box Culverts: Physical Modelling." *Journal of Hydro-environment Research*, IAHR, Vol. 19, pp. 214–223 (DOI: 10.1016/j.jher.2017.03.001).

Title:	Duboulay's rainbowfish swimming upstream in a box culvert barrel flume with asymmetrical large roughness
Filename:	CIMG1410.mov
Duration:	5 min 26 s
Camera details:	Casio Exilim EX-10, movie mode set at 240 fps (replay 30 fps), 512 × 384 pixels
Description:	Duboulay's rainbowfish (*Melanotaenia duboulayi*) swimming upstream in the right bottom corner of a culvert barrel flume (12 m long, 0.5 m wide) equipped with asymmetrical large roughness for q = 0.052 m²/s, V_{mean} = 0.54 m/s, θ = 0 – fish characteristics: L_f = 80 mm, m_f = 4.4.1 g (same fish as for movie CIMG1409.mov) – the fish swims next to the left rough sidewall, in an almost stationary manner, using the corner roughness arrangement to shelter
Relevant book sections:	Chapter 4
Reference:	Wang, H., Chanson, H., Kern, P., and Franklin, C. (2016). "Culvert Hydrodynamics to enhance Upstream Fish Passage: Fish Response to Turbulence." *Proceedings of 20th Australasian Fluid Mechanics Conference*, Australasian Fluid Mechanics Society, G. Ivey, T. Zhou, N. Jones, S. Draper Editors, Perth WA, Australia, 5–8 December, Paper 682, 4 pages.
	Wang, H., Uys, W., and Chanson, H. (2018). "Alternative Mitigation Measures for Fish Passage in Standard Box Culverts: Physical Modelling." *Journal of Hydro-environment Research*, IAHR, Vol. 19, pp. 214–223 (DOI: 10.1016/j.jher.2017.03.001).

F.3.4 Smooth culvert barrel channel with plain triangular corner baffles on one side only

Title:	Juvenile silver perch swimming upstream in a smooth box culvert barrel flume and negotiating a small triangular corner baffle
Filename:	CIMG2791_20170310_3.mov
Duration:	22 s
Camera details:	Casio Exilim EX-10, movie mode set at 240 fps (replay 30 fps), 512 × 384 pixels
Description:	Juvenile silver perch (*Bidyanus bidyanus*) swimming upstream next to the left bottom corner of a smooth culvert barrel flume equipped with triangular bottom corner baffle (L_b = 0.67 m, h_b = 0.067 m, left side only) for q = 0.112 m²/s, V_{mean} = 0.68 m/s, θ = 0 – fish characteristics: L_f = 62 mm, m_f = 2.6 g – fish negotiating successfully the triangular corner baffle, swimming near the base of the baffle

Relevant book sections:	Chapter 7 and Appendix E
Reference:	Cabonce, J., Fernando, R., Wang, H., and Chanson, H. (2019). "Using Small Triangular Baffles to Facilitate Upstream Fish Passage in Standard Box Culverts." *Environmental Fluid Mechanics*, Vol. 19, No. 1, pp. 157–179 (DOI: 10.1007/s10652–018–9604-x)

Title:	Juvenile silver perch swimming in a smooth box culvert barrel flume in the wake of a small triangular corner baffle
Filename:	CIMG2796_20170_4_short.avi
Duration:	31 s
Camera details:	Casio Exilim EX-10, movie mode set at 240 fps (replay 30 fps), 512 × 384 pixels
Description:	Juvenile silver perch (*Bidyanus bidyanus*) swimming upstream next to the left bottom corner of a smooth culvert barrel flume equipped with triangular bottom corner baffle ($L_b = 0.67$ m, $h_b = 0.067$ m, left side only) for $q = 0.112$ m^2/s, $V_{mean} = 0.68$ m/s, $\theta = 0$ – fish characteristics: $L_f = 42$ mm, $m_f = 1.4$ g – fish resting in the wake of (i.e. downstream of) the small triangular bottom corner baffle
Relevant book sections:	Chapter 7 and Appendix E
Reference:	Cabonce, J., Fernando, R., Wang, H., and Chanson, H. (2019). "Using Small Triangular Baffles to Facilitate Upstream Fish Passage in Standard Box Culverts." *Environmental Fluid Mechanics*, Vol. 19, No. 1, pp. 157–179 (DOI: 10.1007/s10652–018–9604-x)

Title:	Juvenile silver perch swimming in a smooth box culvert barrel flume immediately upstream of a small triangular corner baffle
Filename:	CIMG2835_20170317_1_short.avi
Duration:	29 s
Camera details:	Casio Exilim EX-10, movie mode set at 240 fps (replay 30 fps), 512 × 384 pixels
Description:	Juvenile silver perch (*Bidyanus bidyanus*) swimming upstream next to the left bottom corner of a smooth culvert barrel flume equipped with triangular bottom corner baffle ($L_b = 0.67$ m, $h_b = 0.067$ m, left side only) for $q = 0.112$ m^2/s, $V_{mean} = 0.68$ m/s, $\theta = 0$ – fish characteristics: $L_f = 65$ mm, $m_f = 3.1$ g – fish resting immediately upstream of (i.e. upstream of) the small triangular bottom corner baffle, in the stagnation region
Relevant book sections:	Chapter 7 and Appendix E
Reference:	Cabonce, J., Fernando, R., Wang, H., and Chanson, H. (2019). "Using Small Triangular Baffles to Facilitate Upstream Fish Passage in Standard Box Culverts." *Environmental Fluid Mechanics*, Vol. 19, No. 1, pp. 157–179 (DOI: 10.1007/s10652–018–9604-x)

Title:	Juvenile silver perch swimming upstream in a smooth box culvert barrel flume and negotiating a small triangular corner baffle
Filename:	CIMG2874_20170322_7.mov
Duration:	6 s
Camera details:	Casio Exilim EX-10, movie mode set at 240 fps (replay 30 fps), 512 × 384 pixels

Description:	Juvenile silver perch (*Bidyanus bidyanus*) swimming upstream next to the left bottom corner of a smooth culvert barrel flume equipped with triangular bottom corner baffle (L_b = 0.67 m, h_b = 0.067 m, left side only) for q = 0.112 m²/s, V_{mean} = 0.68 m/s, θ = 0 – fish characteristics: L_f = 44 mm, m_f = 0.8 g – fish negotiating successfully the triangular corner baffle, swimming mid-height of the baffle
Relevant book sections:	Chapter 7 and Appendix E
Reference:	Cabonce, J., Fernando, R., Wang, H., and Chanson, H. (2019). "Using Small Triangular Baffles to Facilitate Upstream Fish Passage in Standard Box Culverts." *Environmental Fluid Mechanics*, Vol. 19, No. 1, pp. 157–179 (DOI: 10.1007/s10652–018–9604-x)

Title:	Juvenile silver perch swimming upstream in a smooth box culvert barrel flume and negotiating a small triangular corner baffle
Filename:	CIMG2822.mov
Duration:	28 s
Camera details:	Casio Exilim EX-10, movie mode set at 240 fps (replay 30 fps), 512 × 384 pixels
Description:	Juvenile silver perch (*Bidyanus bidyanus*) swimming upstream next to the left bottom corner of a smooth culvert barrel flume equipped with triangular bottom corner baffle (L_b = 0.67 m, h_b = 0.133 m, left side only) for q = 0.112 m²/s, V_{mean} = 0.64 m/s, θ = 0 – fish characteristics: L_f = 75 mm, m_f = 4.5 g – fish negotiating successfully the triangular corner baffle, swimming next to the bed in a relatively "lazy" motion, before resting upstream of the baffle in the stagnation region
Relevant book sections:	Chapter 7 and Appendix E
Reference:	Cabonce, J., Fernando, R., Wang, H., and Chanson, H. (2019). "Using Small Triangular Baffles to Facilitate Upstream Fish Passage in Standard Box Culverts." *Environmental Fluid Mechanics*, Vol. 19, No. 1, pp. 157–179 (DOI: 10.1007/s10652–018–9604-x)

Title:	Juvenile silver perch in a smooth box culvert barrel flume trapped in the near-wake of a baffle
Filename:	CIMG2830.mov
Duration:	17 mion 52 s
Camera details:	Casio Exilim EX-10, movie mode set at 240 fps (replay 30 fps), 512 × 384 pixels
Description:	Juvenile silver perch (*Bidyanus bidyanus*) swimming upstream next to the left bottom corner of a smooth culvert barrel flume equipped with triangular bottom corner baffle (L_b = 0.67 m, h_b = 0.133 m, left side only) for q = 0.112 m²/s, V_{mean} = 0.64 m/s, = 0 – fish characteristics: L_f = 59 mm, m_f = 2.2 g – fish trapped in the near wake immediately downstream of baffle, and subjected to vortex shedding actions
Relevant book sections:	Chapter 7 and Appendix E
Reference:	Cabonce, J., Fernando, R., Wang, H., and Chanson, H. (2019). "Using Small Triangular Baffles to Facilitate Upstream Fish Passage in Standard Box Culverts." *Environmental Fluid Mechanics*, Vol. 19, No. 1, pp. 157–179 (DOI: 10.1007/s10652–018–9604-x)

Table F.1 Fish swimming in 12 m long and 0.5 m wide box culvert barrel channels

Reference	Q (m³/s)	Bcell (m)	d (¹) (m)	Vmean (¹) (m/s)	T (°C)	Fish species
(1)	(2)	(3)	(4)	(5)	(6)	(7)
Wang et al. (2016a)						
Smooth channel	0.0261	0.50	0.123	0.424	24.5	Duboulay's rainbowfish (*Melanotaenia duboulayi*)
Rough bed and smooth sidewalls	0.0261	0.50	0.133	0.392	±0.5	Duboulay's rainbowfish (*Melanotaenia duboulayi*)
Rough bed and rough left sidewall	0.0261	0.478	0.129	0.424		Juvenile silver perch (*Bidyanus bidyanus*)
						Duboulay's rainbowfish (*Melanotaenia duboulayi*)
Cabonce et al. (2017, 2018, 2019)						
Smooth channel	0.0556	0.50	0.162	0.686	24.5	Juvenile silver perch (*Bidyanus bidyanus*)
Medium baffles (h_b = 0.067 m, L_b = 0.67 m)	0.0556	0.50	0.1625	0.684	±0.5	
Large baffles (h_b = 0.133 m, L_b = 0.67 m)	0.0556	0.50	0.173	0.643		
Large baffles (h_b = 0.133 m, L_b = 0.67 m) with perforation (∅ = 13 mm)	0.0556	0.50	0.173	0.643		

Notes: B_{cell}: internal barrel channel width; d: water depth; h_b: isosceles triangular baffle size; L_b: isosceles triangular baffle spacing; Q: water discharge; T: water temperature; V_{mean}: bulk velocity; l: values recorded 8 m downstream of the flume's entrance. All experiments conducted in a horizontal channel (θ = 0).

References

Titles of textbook references are italicized to emphasize their importance.

Adrian, R.J. & Marusic, I. (2012) Coherent structures in flow over hydraulic engineering surfaces. *Journal of Hydraulic Research*, IAHR, 50(5), 451–464. DOI: 10.1080/00221686.2012.729540.

Alexander, R.M. (1982) *Locomotion of Animals*. Blackie, Glasgow, UK. 163 pages.

Allen, G.R., Midgley, S.H. & Allen, M. (2002) *Field Guide to the Freshwater Fishes of Australia*. Western Australia Museum, Perth, Australia. 394 pages.

Anderson, G.B., Freeman, M.C., Freeman, B.J., Straight, C.A., Hagler, M.M. & Peterson, J.T. (2012) Dealing with uncertainty when assessing fish passage through culvert road crossings. *Environmental Management*, 50, 462–477.

ANSYS® Academic Research, Release 18.0 (2017) Help System, ANSYS FLUENT User's Guide. ANSYS Inc., USA.

Apelt, C.J. (1983) Hydraulics of minimum energy culverts and bridge waterways. *Australian Civil Engineering Transactions*, Institution of Engineers, Australia, CE25(2), 89–95.

Australian Standard (2010) Precast reinforced concrete box culverts: Part 1: Small (not exceeding 1200 mm span and 1200 mm height). Australian Standards AS1597.1–2010. Standard Australia. 53 pages.

Australian Standard (2013) Precast reinforced concrete box culverts: Part 2: Large culverts (exceeding 1200 mm span or 1200 mm height and up to and including 4200mm span and 4200 mm height). Australian Standards AS1597.2–2013. Standard Australia. 73 pages.

Bailly, C. & Comte-Bellot, G. (2015) *Turbulence*. Springer, Heidelberg, Germany. 360 pages.

Bakhmeteff, B.A. (1912) *O Neravnomernom Dwijenii Jidkosti v Otkrytom Rusle* (Varied Flow in Open Channel). St Petersburg, Russia (in Russian).

Bakhmeteff, B.A. & Matzke, A.E. (1936) The hydraulic jump in terms of dynamic similarity. *Transactions*, ASCE, 101, 630–647. Discussion: 101, 648–680.

Ball, J., Babister, M., Nathan, R., Weeks, W., Weinmann, E., Retallick, M. & Testoni, I. (2016) *Australian Rainfall and Runoff: A Guide to Flood Estimation*. Commonwealth of Australia, Canberra ACT, Australia.

Barré de Saint-Venant, A.J.C. (1871a) Théorie du Mouvement Non Permanent des Eaux, avec Application aux Crues des Rivières et à l'Introduction des Marées dans leur Lit. *Comptes Rendus des séances de l'Académie des Sciences*, Paris, France, 73(4), 147–154 (in French).

Barré de Saint-Venant, A.J.C. (1871b) Théorie et Equations Générales du Mouvement Non Permanent des Eaux, avec Application aux Crues des Rivières et à l'Introduction des Marées dans leur Lit (2ème Note). *Comptes Rendus des séances de l'Académie des Sciences*, Paris, France, Séance, 17 July, 73, 237–240 (in French).

Bates, K., Barnard, B., Heiner, B., Klavas, J.P. & Powers, P.D. (2003) *Design of Road Culverts for Fish Passage*. Washington Department of Fish and Wildlife, Olympia, Washington. 111 pages.

Bazin, H. (1865) Recherches Expérimentales sur la Propagation des Ondes (Experimental research on wave propagation). *Mémoires présentés par divers savants à l'Académie des Sciences*, Paris, France, 19, 495–644 (in French).

Behlke, C.E., Kane, D.L., McLeen, R.F. & Travis, M.T. (1991) Fundamentals of culvert design for passage of weak-swimming fish. Report FHW A-AK-RD-90–10, Department of Transportation and Public Facilities, State of Alaska, Fairbanks, USA. 178 pages.

Belanger, J.B. (1828) *Essai sur la Solution Numérique de quelques Problèmes Relatifs au Mouvement Permanent des Eaux Courantes* (Essay on the Numerical Solution of Some Problems relative to Steady Flow of Water). Carilian-Goeury, Paris, France. 38 pages & 5 tables (in French).

Bélanger, J.B. (1841) *Notes sur l'Hydraulique* (Notes on Hydraulic Engineering). Ecole Royale des Ponts et Chaussées, Paris, France, session 1841–1842. 223 pages (in French).

Blake, R.E. (1983) *Fish Locomotion*. Cambridge University Press, Cambridge, UK. 208 pages.

Blank, M.D. (2008) *Advanced Studies of Fish Passage through Culverts: 1-D and 3-D Hydraulic Modelling of Velocity, Fish Energy Expenditure, and a New Barrier Assessment Method*. Ph.D. Thesis, Montana State University, Department of Civil Engineering. 231 pages.

Boussinesq, J.V. (1871) Sur le Mouvement Permanent Varié de l'Eau dans les Tuyaux de Conduite et dans les Canaux Découverts (On the steady varied flow of water in conduits and open channels). *Comptes Rendus des séances de l'Académie des Sciences*, Paris, France, 73(19), 101–105 (in French).

Boussinesq, J.V. (1877) Essai sur la théorie des eaux courantes. *Mémoires présentés par divers savants à l'Académie des Sciences*, 23(1), 1–680 (in French).

Boussinesq, J.V. (1896) Théorie de l'Ecoulement Tourbillonnant et Tumultueux des Liquides dans les Lits Rectilignes à Grande Section (Tuyaux de Conduite et Canaux Découverts) quand cet Ecoulement s'est régularisé en un Régime Uniforme, c'est-à-dire, moyennement pareil à travers toutes les Sections Normales du Lit (Theory of turbulent and tumultuous flow of liquids in prismatic channels of large cross-sections (pipes and open channels) when the flow is uniform, i.e., constant in average at each cross-section along the flow direction). *Comptes Rendus des séances de l'Académie des Sciences*, Paris, France, 122, 1290–1295 (in French).

Boys, P.F.D. du (1879) Etude du Régime et de l'Action exercée par les Eaux sur un Lit à Fond de Graviers indéfiniment affouillable (Study of flow regime and force exerted on a gravel bed of infinite depth). *Ann. Ponts et Chaussées*, Paris, France, série 5, 19, 141–195 (in French).

Bradshaw, P. (1971) *An Introduction to Turbulence and Its Measurement*. Pergamon Press, Oxford, UK, the Commonwealth and International Library of Science and Technology Engineering and Liberal Studies, Thermodynamics and Fluid Mechanics Division. 218 pages.

Bradshaw, P. (1987) Turbulent secondary flows. *Annual Review of Fluid Mechanics*, 19, 53–74.

Bresse, J.A. (1860) *Cours de Mécanique Appliquée Professé à l'Ecole des Ponts et Chaussées* (Course in Applied Mechanics Lectured at the Pont-et-Chaussées Engineering School). Mallet-Bachelier, Paris, France (in French).

Briggs, A.S. & Galarowicz, T.L. (2013) Fish passage through culverts in central Michigan warmwater streams. *North American Journal of Fisheries Management*, 33, 652–664.

Brown, G.O. (2002) Henry Darcy and the making of a law. *Water Research Resources*, 38(7), Paper 11, 11–1 to 11–12.

Buat, P.L.G. du (1779) *Principes d'Hydraulique, vérifiés par un grand nombre d'expériences faites par ordre du gouvernement* (Hydraulic Principles, Verified by a Large Number of Experiments). Imprimerie de Monsieur, Paris, France, 1st edition (in French); 2nd edition: 1786, Paris, France, 2 volumes; 3rd edition: 1816, Paris, France, 3 volumes.

Buckingham, E. (1914) On physically similar systems: Illustrations of the use of dimensional equations. *Physical Review*, 4(4), 345–376.

Cabonce, J., Fernando, R., Wang, H. & Chanson, H. (2017) Using triangular baffles to facilitate upstream fish passage in box culverts: Physical modelling. Hydraulic Model Report No. CH107/17, School of Civil Engineering, The University of Queensland, Brisbane, Australia. 130 pages (ISBN 978-1-74272-186-6).

Cabonce, J., Wang, H. & Chanson, H. (2018) Ventilated corner baffles to assist upstream passage of small-bodied fish in box culverts. *Journal of Irrigation and Drainage Engineering*, ASCE, 144(8), Paper 0418020, 8 pages. DOI: 10.1061/(ASCE)IR.1943-4774.0001329.

Cabonce, J., Fernando, R., Wang, H. & Chanson, H. (2019) Using small triangular baffles to facilitate upstream fish passage in standard box culverts. *Environmental Fluid Mechanics*, 19(1), 157–179. DOI: 10.1007/s10652-018-9604-x.

Cahoon, J.E., McMahon, T., Solcz, A., Blank, M.D. & Stein, O. (2007) Fish passage in Montana culverts: Phase II: Passage goals. Report FHWA/MT-07–010/8181, Montana Department of Transportation and US Department of Transportation, Federal Highway Administration. 61 pages.

Chanson, H. (1999a) Culvert design. In: *The Hydraulics of Open Channel Flow: An Introduction*. Butterworth-Heinemann, H. Chanson, London, UK, 1st edition. pp. 365–401 (ISBN 0-340-74067-1).

Chanson, H. (1999b) Physical modelling of hydraulics. In: *The Hydraulics of Open Channel Flow: An Introduction*. Butterworth-Heinemann, H. Chanson, London, UK, 1st edition. pp. 261–283 (ISBN 0-340-74067-1).

Chanson, H. (2004) *The Hydraulics of Open Channel Flow: An Introduction*. Butterworth-Heinemann, Oxford, UK, 2nd edition. 630 pages (ISBN 978-0-7506-5978-9).

Chanson, H. (2006) Minimum specific energy and critical flow conditions in open channels. *Journal of Irrigation and Drainage Engineering*, ASCE, 132(5), 498–502. Doi: 10.1061/(ASCE)0733-9437(2006)132:5(498).

Chanson, H. (2008) Minimum specific energy and critical flow conditions in open channels: Closure. *Journal of Irrigation and Drainage Engineering*, ASCE, 134(6), 883–887. DOI: 10.1061/(ASCE)0733-9437(2008)134:6(883).

Chanson, H. (2009a) Development of the Bélanger equation and backwater equation by Jean-Baptiste Bélanger (1828). *Journal of Hydraulic Engineering*, ASCE, 135(3), 159–163. DOI: 10.1061/(ASCE)0733-9429(2009)135:3(159).

Chanson, H. (2009b) Turbulent air-water flows in hydraulic structures: Dynamic similarity and scale effects. *Environmental Fluid Mechanics*, 9(2), 125–142. DOI: 10.1007/s10652-008-9078-3.

Chanson, H. (2014) *Applied Hydrodynamics: An Introduction*. CRC Press, Taylor & Francis Group, Leiden, The Netherlands. 448 pages & 21 video movies (ISBN 978-1-138-00093-3).

Chanson, H. (2019) Utilising the boundary layer to help restore the connectivity of fish habitats and populations: An engineering discussion. *Ecological Engineering*, 141, Paper 105613, 5 pages. DOI: 10.1016/j.ecoleng.2019.105613.

Chanson, H. & Leng, X. (2018) On the development of hydraulic engineering guidelines for fish-friendly standard box culverts, with a focus on small-body fish. In: *Civil Engineering Research Bulletin No. 25*. School of Civil Engineering, The University of Queensland, Brisbane, Australia. 79 pages (ISBN 978-1-74272-208-5).

Chanson, H. & Leng, X. (2019) There is something fishy about turbulence: Why novel hydraulic engineering guidelines can assist the upstream passage of small-bodied fish species in standard box culverts. In: *Civil Engineering Research Bulletin No. 26*. School of Civil Engineering, The University of Queensland, Brisbane, Australia. 224 pages (ISBN 798-1-74272-234-4).

Chanson, H. & Montes, J.S. (1995) Characteristics of undular hydraulic jumps: Experimental apparatus and flow patterns. *Journal of Hydraulic Engineering*, ASCE, 121(2), 129–144. DOI: 10.1061/(ASCE)0733-9429(1995)121:2(129).

Chorda, J., Larinier, M. & Font, S. (1995) Le Franchissement par les Poissons Migrateurs des Buses et Autres Ouvrages de Rétablissement des Ecoulements Naturels lors des Aménagements Routiers et Autoroutes. Etude Expérimentale. Rapport HYDRE n°159 – GHAAPPE n°95–03, Groupe d'Hydraulique Appliquée aux Aménagements Piscicoles et à la Protection de l'Environnement, Service d'Etudes Techniques des Routes et Autoroutes, Toulouse, France. 116 pages (in French).

Chow, V.T. (1959) *Open Channel Hydraulics*. McGraw-Hill, New York, USA.

CIRIA (1987) Protection and provision for safe overtopping of dams and flood banks. CIRIA Project Report, Construction Industry Research and Information Association, London, UK. 126 pages.

Clanet, C. (2013) *Sports Physics*. Editions de l'Ecole Polytechnique, Palaiseau, France. 633 pages.

Clarke, S.J., Mostyn, G. & Shen, B. (2019) Collapse of the Old Pacific Highway, Piles Creek, Somersby. *Australian Geomechanics*, 64(3), 81–97.

Colavecchia, M., Katopodis, C., Goosney, R., Scruton, D.A. & McKinley, R.S. (1998) Measurement of burst swimming performance in wild Atlantic salmon (*Salmo Salar L.*) using digital telemetry. *Regulated Rivers Research and Management*, 14, 41–51.

Concrete Pipe Association of Australasia (1991) *Hydraulics of Precast Concrete Conduits*. Jenkin Buxton Printers, Australia, 3rd edition. 72 pages.

Concrete Pipe Association of Australasia (2012) *Hydraulics of Precast Concrete Conduits*. CPAA Design Manual, Australia, 5th edition. 64 pages.

Coriolis, G.G. (1836) Sur l'établissement de la formule qui donne la figure des remous et sur la correction qu'on doit introduire pour tenir compte des différences de vitesses dans les divers points d'une même section d'un courant (On the establishment of the formula giving the backwater curves and on the correction to be introduced to take into account the velocity differences at various points in a cross-section of a stream). *Annales des Ponts et Chaussées*, 1st Semester, Series 1, 11, 314–335 (in French).

Cotel, A.J., and Webb, P.W. (2015). "Living in a Turbulent World – A New Conceptual Framework for the Interactions of Fish and Eddies." *Integrative and Comparative Biology*, Vol. 55, No. 4, pp. 662–672.

Cottman, N.H., Porter, K.F., Tiller, J.E. & Tonkin, B.C. (1980) A commentary and bibliography on the hydraulics of culvert design. ARRB Internal Report AIR 806-4, Australian Road Research Board, Vermont South VIC. 78 pages.

Counsilman, J.E. (1968) *The Science of Swimming*. Prentice-Hall, USA. 457 pages.

Courret, D. (2014) Petits Ouvrages Hydrauliques et continuite pisicole. Equipement des OH existants: Principes de choix et de dimensionnement des dispositifes. *Presentation*, Nancy. 23 pages.

Crowder, D.W. & Diplas, P. (2000) Evaluating spatially explicit metrics of stream energy gradients using hydrodynamics model simulations. *Canadian Journal of Fisheries and Aquatic Sciences*, 57, 1497–1507.

Darcy, H.P.G. (1856) *The Public Fountains of the City of Dijon: Exposition and Application of Principles to Follow and Formulas to Use in Questions of Water Distribution*. Kendal/Hunt Publ., Dubuque, Iowa, English Translation by P. Bobeck.

Darcy, H.P.G. (1858) Recherches Expérimentales relatives aux Mouvements de l'Eau dans les Tuyaux (Experimental research on the motion of water in pipes). *Mémoires Présentés à l'Académie des Sciences de l'Institut de France*, 14, 141 (in French).

Darcy, H.P.G. & Bazin, H. (1865) *Recherches Hydrauliques* (Hydraulic Research). Imprimerie Impériales, Paris, France, Parties 1ère et 2ème (in French).

Duguay, J., Foster, B., Lacey, J. & Castro-Santos, T. (2018) Sediment infilling benefits rainbow trout passage in a baffled channel. *Ecological Engineering*, 125, 38–49. DOI: 10.1016/j.ecoleng.2018.10.003.

Dupuit, A.J.E. (1848) *Etudes Théoriques et Pratiques sur le Mouvement des Eaux Courantes* (Theoretical and Practical Studies on Flow of Water). Dunod, Paris, France (in French).

DWA (2014) Fischaufstiegsanlagen und fischpassierbare Bauwerke – Gestaltung, Bemessung, Qualitätssicherung. Merkblatt DWA-M 509, Deutsche Vereinigung für Wasserwirtschaft, Abwasser und Abfall e. V., Hennef, Germany. 334 pages (ISBN 978-3-942964-91-3).

Dynesius, M. & Nilsson, C. (1994) Fragmentation and flow regulation of river systems in the northern third of the world. *Science*, 266, 753–762.

Einstein, H.A. & LI, H. (1956) Secondary currents in straight channels. *Transactions*, AGU, 39(6), 1085–1088.

Eloy, C. (2012) Optimal Strouhal number for swimming animals. *Journal of Fluids and Structures*, 30, 205–218.

Fairfull, S. & Witheridge, G. (2003) *Why Do Fish Need to Cross the Road? Fish Passage Requirements for Waterway Crossings*. NSW Fisheries, Cronulla NSW, Australia. 14 pages.

Fawer, C. (1937) *Etude de Quelques Ecoulements Permanents à Filets Courbes* (Study of Some Steady Flows with Curved Streamlines). Thesis, Lausanne, Switzerland, Imprimerie La Concorde. 127 pages (in French).

Feynman, R., Leighton, R.B. & Sands, M. (1964) *The Feynman Lectures on Physics*. Addison-Wesley, Reading, MA, 3 volumes.

Foss, J.F., Panton, R. & Yarin, A. (2007) Nondimensional representation of the boundary-value problem. In: Tropea, C., Yarin, A.L. & Foss, J.F. (eds.) *Springer Handbook of Experimental Fluid Mechanics*. Springer, Germany. Part A, Chapter 2. pp. 33–82.

Freire, R., Sailema, C. & Chanson, H. (2018) On ventilated corner baffles for box culvert barrel: A physical investigation. Hydraulic Model Report No. CH112/18, School of Civil Engineering, The University of Queensland, Brisbane, Australia. 91 pages.

Gardner, A. (2006) *Fish Passage through Road Culverts*. Master of Science Thesis, North Carolina State University, USA. 103 pages.

Gauckler, P.G. (1867) Etudes Théoriques et Pratiques sur l'Ecoulement et le Mouvement des Eaux (Theoretical and practical studies of the flow and motion of waters). *Comptes Rendues de l'Académie des Sciences*, Paris, France, Tome 64, 818–822 (in French).

Gessner, F.B. (1973) The origin of secondary flow in turbulent flow along a corner. *Journal of Fluid Mechanics*, 58(1), 1–25.

Goettel, M.T., Atkinson, J.F. & Bennett, S.J. (2015) Behavior of western blacknose dace in a turbulence modified flow field. *Ecological Engineering*, 74, 230–240.

Goodrich, H.R., Watson, J.R., Cramp, R.L, Gordos, M. & Franklin, C.E. (2018) Making culverts great again: Efficacy of a common culvert remediation strategy across sympatric fish species. *Ecological Engineering*, 116, 143–153.

Guiny, E., Ervine, D.A. & Amstrong, J.D. (2005) Hydraulic and biological aspects of fish passes for Atlantic salmon. *Journal of Hydraulic Engineering*, ASCE, 131(7), 542–553. DOI: 10.1061/(ASCE)0733-9429~2005!131:7(542).

Haro, A., Castro-Santos, T., Noreika, J. & Odeh, M. (2004) Swimming performance of upstream migrant fishes in open-channel flow: A new approach to predicting passage through velocity barriers. *Canadian Journal of Fisheries and Aquatic Sciences*, 61, 1590–1601.

Hee, M. (1969) Hydraulics of culvert design including constant energy concept. *Proc. 20th Conf. of Local Authority Engineers*, Dept. of Local Govt, Queensland, Australia, Paper 9. pp. 1–27.

Henderson, F.M. (1966) *Open Channel Flow*. MacMillan Company, New York, USA.

Herr, L.A. & Bossy, H.G. (1965) Capacity charts for the hydraulic design of highway culverts. Hydraulic Eng. Circular, US Dept. of Transportation, Federal Highway Admin., Washington DC, USA, HEC No. 10, March.

Hirt, C. & Nichols, B. (1981) Volume of Fluid (VOF) method for the dynamics of free boundaries. *Journal of Computational Physics*, 39(1), 201–225.

Hong, J.R., Katz, J. & Schultz, M.P. (2011) Near-wall turbulence statistics and flow structures over three-dimensional roughness in a turbulent channel flow. *Journal of Fluid Mechanics*, 667, 1–37. DOI: 10.1017/S0022112010003988.

Hotchkiss, R.H. (2002) Turbulence investigation and reproduction for assisting downstream migrating juvenile salmonids, Part I. BPA Report DOE/BP-00004633-I, Bonneville Power Administration, Portland, Oregon. 138 pages.

Hotchkiss, R.H. & Frei, C.M. (2007) Design for fish passage at roadway-stream crossings: Synthesis report. Publication No. FHWA-HIF-07-033, Federal Highway Administration, US Department of Transportation. 280 pages.

Howe, J.W. (1949) Flow measurement. *Proc 4th Hydraulic Conf.*, Iowa Institute of Hydraulic Research, John Wiley & Sons Publ., June, Rouse, H. (ed.). pp. 177–229.

Jensen, K.M. (2014) *Velocity Reduction Factors in Near Boundary Flow and the Effect on Fish Passage through Culverts*. Master of Science Thesis, Brigham Young University, USA. 44 pages.

Katopodis, C. & Gervais, R. (2016) *Fish Swimming Performance Database and Analyses*. DFO CSAS Research Document No. 2016/002, Canadian Science Advisory Secretariat, Fisheries and Oceans Canada, Ottawa, Canada. 550 pages.

Kemp, P. (2012) Bridging the gap between fish behaviour, performance and hydrodynamics: An eco-hydraulics approach to fish passage research. *River Research and Applications*, 28, 403–406. DOI: 10.1002/rra.1599.

Kern, P., Cramp, R., Gordos, M.A., Watson, J. & Franklin, C. (2018) Measuring U_{crit} and endurance: Equipment choice influences estimates of fish swimming performance. *Journal of Fish Biology*, 92, 237–247.

Khodier, M.A. & Tullis, B.P. (2014) Fish passage behavior for severe hydraulic conditions in baffled culverts. *Journal of Hydraulic Engineering*, ASCE, 140(3), 322–327. DOI: 10.1061/(ASCE) HY.1943-7900.0000831.

Khodier, M.A. & Tullis, B.P. (2018) Experimental and computational comparison of baffled-culvert hydrodynamics for fish passage. *Journal of Applied Water Engineering and Research*, 6(3), 191–199. DOI: 10.1080/23249676.2017.1287018.

Kilgore, R.T., Bergendahl, B.S. & Hotchkiss, R.H. (2010) *Culvert Design for Aquatic Passage*. Hydraulic Engineering Circular Number No. 26. Federal Highway Administration Publication No. FHWA-HIF-11-008. 234 pages.

Kobus, H. (1984) Scale effects in modelling hydraulic structures. *Proceedings of the International Symposium on Scale Effects in Modelling Hydraulic Structures*, IAHR, Esslingen, Germany.

Kolmogorov, S.V. & Duplishcheva, O.A. (1992) Active drag, useful mechanical power output and hydrodynamic force coefficient in different swimming strokes at maximal velocity. *Journal of Biomechanics*, 25(3), 311–318.

Lacey, R.W.J., Neary, V.S., Liao, J.C., Enders, E.C. & Tritico, H.M. (2012) The IPOS framework: Linking fish swimming performances in altered flows from laboratory experiments to rivers. *River Research and Applications*, 28(4), 429–443. DOI: 10.1002/rra.1584.

Larinier, M. (2002) Fish passage through culverts, rock weirs and estuarine obstructions. *Bulletin Français de Pêche et Pisciculture*, 364(Supplement), 119–134.

Launder, B.E. & Spalding, D.B. (1974) The numerical computation of turbulent flows. *Computer Methods in Applied Mechanics and Engineering*, 3(2), 269–289.

Leng, X. & Chanson, H. (2018) Modelling low velocity zones in box culverts to assist fish passage. *Proceedings of 21st Australasian Fluid Mechanics Conference*, 10–13 December, Adelaide, Australia, Lau, T.C.W. & Kelso, R.M. (eds.), Paper 547. 4 pages.

Leng, X., Chanson, H., Gordos, M. & Riches, M. (2019) Developing cost-effective design guidelines for fish-friendly box culverts, with a focus on small fish. *Environmental Management*, 63(6), 747–758 & Supplementary material (7 pages). DOI: 10.1007/s00267-019-01167-6 (ISSN 0364-152X).

Lesieur, M. (1994) *La Turbulence* (The Turbulence). Presses Universitaires de Grenoble, Collection Grenoble Sciences, France. 262 pages (in French).

Levi, E. (1965) Longitudinal streaking in liquid currents. *Journal of Hydraulic Research*, IAHR, 3(2), 25–39.

Liggett, J.A. (1994) *Fluid Mechanics*. McGraw-Hill, New York, USA.

Liggett, J.A., Chiu, C.L. & Miao, L.S. (1965) Secondary currents in a corner. *Journal of Hydraulic Division*, ASCE, 91(HY6), 99–117.

Lighthill, M.J. (1969) Hydromechanics of aquatic animal propulsion. *Annual Review of Fluid Mechanics*, 1, 413–446.

Lighthill, M.J. (1975) *Mathematical Biofluiddynamics*. Society for Industrial and Applied Mathematics, Philadelphia, USA. 278 pages.

Lintermans, M. (2013) Recovering threatened freshwater fish in Australia. *Marine and Freshwater Research*, 64, iii–vi. DOI: 10.1071/MFv64n9_IN.

Liu, M.M., Rajaratnam, N. & Zhu, D.Z. (2006) Mean flow and turbulence structure in vertical slot fishways. *Journal of Hydraulic Engineering*, ASCE, 139(4), 424–432.

Lupandin, A.I. (2005) Effect of flow turbulence on swimming speed of fish. *Biology Bulletin*, 32(5), 461–466.

Macintosh, J.C. (1990) *Hydraulic Characteristics in Channels of Complex Cross-Section*. Ph.D. Thesis, School of Civil Engineering, The University of Queensland, Brisbane, Australia. 505 pages.

Maddock, I., Harby, A., Kemp, P. & Wood, P. (2013) Research needs, challenges and the future of ecohydraulics research. In: Maddock, I., Harby, A., Kemp, P. & Wood, P. (eds.) *Ecohydraulics: An Integrated Approach*. John Wiley, USA, Chapter 25. pp. 431–436.

Manning, R. (1890) *On the Flow of Water in Open Channels and Pipes*. Instn of Civil Engineers of Ireland, Ireland.

Milt, A.W., Diebel, M.W., Doran, P.J., Ferris, M.C., Herbert, M., Khoury, M.L., Moody, A.T., Neeson, T.M., Ross, J., Treska, T., O'Hanley, J.R., Walter, L., Wangen, A.R., Yacobson, E. & McIntyre, P.B. (2018) Minimizing opportunity costs to aquatic connectivity restoration while controlling an invasive species. *Conservation Biology*, 32(4), 894–904. DOI: 10.1111/cobi.13105.

Monk, S.K. & Hotchkiss, R.H. (2012) Culvert roughness elements for native utah fish passage: Phase II. Report No. UT-12.09, Utah Department of Transportation – Research Division, USA. 47 pages.

Montes, J.S. (1998) *Hydraulics of Open Channel Flow*. ASCE Press, New York, USA. 697 pages.

Morvan, H., Knight, D., Wright, N., Tang, X. & Crossley, A. (2008) The concept of roughness in fluvial hydraulics and its formulation in 1D, 2D and 3D numerical simulation models. *Journal of Hydraulic Research*, 46(2), 191–208.

Naot, D. & Rodi, W. (1982) Numerical simulations of secondary currents in channel flow. *Journal of Hydraulic Division*, ASCE, 108(HY8), 948–968.

Navier, M. (1823) *Mémoire sur les Lois du Mouvement des Fluides* (Memoirs on the Laws of Fluid Motion). Mém. Acad. des Sciences, Paris, France, volume 6. pp. 389–416.

Nezu, I. & Rodi, W. (1985) Experimental study on secondary currents in open channel flow. *Proceedings of the 21st IAHR Congress*, IAHR, Melbourne. pp. 115–119.

Nikora, V.I., Aberle, J., Biggs, B.J.F., Jowett, I.G. & Sykes, J.R.E. (2003) Effects of fish size, time-to-fatigue and turbulence on swimming performance: A case study of galaxias maculatus. *Journal of Fish Biology*, 63, 1365–1382.

Nikuradse, J. (1926) Turbulente Stromung im Innem des rechteckigen offenen Kanals. *Forschungsarbeiten*, Heft, 281, 36–44 (in German).

Novak, P. & Cabelka, J. (1981) *Models in Hydraulic Engineering: Physical Principles and Design Applications*. Pitman Publ., London, UK. 459 pages.

O'Hanley, J.R. (2011) Open rivers: Barrier removal planning and the restoration of free-flowing rivers. *Journal of Environmental Management*, 92, 3112–3120.

Olsen, A. & Tullis, B. (2013) Laboratory study of fish passage and discharge capacity in slip-lined, baffled culverts. *Journal of Hydraulic Engineering*, ASCE, 139(4), 424–432.

Olson, L.J. & Roy, S. (2002) The economics of controlling a stochastic biological invasion. *American Journal of Agricultural Economics*, 84, 1311–1316.

O'Neill, P.L., Nicolaides, D., Honnery, D. & Soiria, J. (2004) Autocorrelation functions and the determination of integral length with reference to experimental and numerical data. *Proceedings of 15th Australasian Fluid Mechanics Conference*, 13–17 December, the University of Sydney, Sydney, Australia. 4 pages (CD-ROM).

Papanicolaou, A.N. & Talebbeydokhti, N. (2002) Discussion of turbulent open-channel flow in circular corrugated culverts. *Journal of Hydraulic Engineering*, ASCE, 128(5), 548–549.

Pavlov, D.S., Lupandin, A.I. & Skorobogatov, M.A. (2000) The effects of flow turbulence on the behavior and distribution of fish. *Journal of Ichthyology*, 40, S232–S261.

Plew, D.R., Nikora, V.I., Larne, S.T., Sykes, J.R.E. & Cooper, G.G. (2007) Fish swimming speed variability at constant flow: Galaxias maculatus. *New Zealand Journal of Marine and Freshwater Research*, 41, 185–195. DOI: 0028-8330/07/4102-0185.

Prandtl, L. (1904) *Über Flussigkeitsbewegung bei sehr kleiner Reibung* (On Fluid Motion with Very Small Friction). Verh. III Intl. Math. Kongr., Heidelberg, Germany (in German) (also NACA Tech. Memo. No. 452, 1928).

QUDM (2016) *Queensland Urban Drainage Manual*. Institute of Public Works Engineering Australasia, Queensland Division, Brisbane, Australia, 4th edition. 391 pages.

Reynolds, O. (1883) An experimental investigation of the circumstances which determine whether the motion of water shall be direct or sinuous, and the laws of resistance in parallel channels. *Philosophical Transactions of Royal Society of London*, 174, 935–982.

Richmond, M.C., Debg, Z., Guensch, G.R., Tritico, H. & Pearson, W.H. (2007) Mean flow and turbulence characteristics of a full-scale spiral corrugated culvert with implications for fish passage. *Ecological Engineering*, 30, 333–340.

Rizzi, A. & Vos, J. (1998) Toward establishing credibility in computational fluid dynamics simulations. *AIAA Journal*, 36(5), 668–675.

Roache, R.L. (1998) *Verification and Validation in Computational Science and Engineering*. Hermosa Publishers, Albuquerque, NM, USA. 446 pages.

Rodi, W. (1995) Impact of Reynolds-average modelling in hydraulics. *Proceedings Mathematical and Physical Sciences*, 451(1941), 141–164.

Rodi, W., Constantinescu, G. & Stoesser, T. (2013) *Large-Eddy Simulation in Hydraulics*. IAHR Monograph. CRC Press, Taylor & Francis Group, Leiden, The Netherlands. 252 pages.

Rouse, H. (1938) *Fluid Mechanics for Hydraulic Engineers*. McGraw-Hill Publ., New York, USA (also Dover Publ., New York, USA, 1961. 422 pages).

Runge, C. (1908) Uber eine Method die partielle Differentialgleichung Δu = constant numerisch zu integrieren. *Zeitschrift der Mathematik und Physik*, 56, 225–232 (in German).

Sanchez, P.X., Leng, X. & Chanson, H. (2018) Fluid dynamics and secondary currents in an asymmetrical rectangular canal with sidewall streamwise rib. Hydraulic Model Report No. CH113/18, School of Civil Engineering, The University of Queensland, Brisbane, Australia. 158 pages (ISBN 978-1-74272-223-8).

Schall, J.D., Thompson, P.L., Zerges, S.M., Kilgore, R.T. & Morris, J.L. (2012) *Hydraulic Design of Highway Culverts*. Hydraulic Design Series No. 5. Federal Highway Administration, Publication No. FHWA-HIF-12-026, USA, 3rd edition. 326 pages.

Schlichting, H. (1979) *Boundary Layer Theory*. McGraw-Hill, New York, USA, 7th edition.

Smits, A.J. & Marusic, I. (2013) Wall-bounded turbulence. *Physics Today*, 66(9), 25–30.

Spiegel, M.R. (1972) *Theory and Problems of Statistics*. McGraw-Hill Inc., New York, USA.

Tominaga, A. & Nezu, I. (1991) Turbulent structure of shear flow with spanwise roughness heterogeneity. *Proc. International Symposium on Environmental Hydraulics*, 16–18 December, Hong Kong, Balkema Publ., Rotterdam. pp. 415–420.

Uijttewaal, W.S.J. (2014) Hydrodynamics of shallow flows: Application to rivers. *Journal of Hydraulic Research*, IAHR, 52(2), 157–172. DOI: 10.1080/00221686.2014.905505.

USBR (1987) *Design of Small Dams: Bureau of Reclamation*. US Department of the Interior, Denver, CO, USA, 3rd edition.

Vaschy, A. (1892) Sur les lois de similitude en physique. *Annales Télégraphiques*, 19, 25–28 (in French).

Videler, J.J. (1993) *Fish Swimming*. Chapman and Hall, London, UK. 260 pages.

Wang, H. & Chanson, H. (2018a) Modelling upstream fish passage in standard box culverts: Interplay between turbulence, fish kinematics, and energetics. *River Research and Applications*, 34(3), 244–252. DOI: 10.1002/rra.3245.

Wang, H. & Chanson, H. (2018b) On upstream fish passage in standard box culverts: Interactions between fish and turbulence. *Journal of Ecohydraulics*, IAHR, 3(1), 18–29. DOI: 10.1080/24705357.2018.1440183.

Wang, H., Chanson, H., Kern, P. & Franklin, C. (2016a) Culvert hydrodynamics to enhance upstream fish passage: Fish response to turbulence. *Proceedings of 20th Australasian Fluid Mechanics Conference*, Australasian Fluid Mechanics Society, Perth WA, Australia, 5–8 December, Ivey, G., Zhou, T., Jones, N. & Draper, S. (eds.), Paper 682. 4 pages.

Wang, H., Beckingham, L.K., Johnson, C.Z., Kiri, U.R. & Chanson, H. (2016b) Interactions between large boundary roughness and high inflow turbulence in open channel: A physical study into turbulence properties to enhance upstream fish migration. Hydraulic Model Report No. CH103/16, School of Civil Engineering, The University of Queensland, Brisbane, Australia. 74 pages (ISBN 978-1-74272-156-9).

Wang, H., Uys, W. & Chanson, H. (2018) Alternative mitigation measures for fish passage in standard box culverts: Physical modelling. *Journal of Hydro-Environment Research*, IAHR, 19, 214–223. DOI: 10.1016/j.jher.2017.03.001.

Wang, X.F. & Wang, L.Z. (2006) Preliminary probe into optimization on propulsion in short distance competitive swimming. *Applied Mathematics and Mechanics*, 27, 117–124. DOI: 10.1007/s10483-006-0115-z.

Warren, M.L., Jr. & Pardew, M.G. (1998) Road crossings as barriers to small-stream fish movement. *Transactions of the American Fisheries Society*, 127, 637–644.

Watson, J.R., Goodrich, H.R., Cramp, R.L., Gordos, M.A., Yan, Y., Ward, P.L., & Franklin, C.E. (2019) Swimming performance traits of twenty-one Australian fish species: a fish passage management tool in a modified freshwater system. *bioRxiv*. DOI: 10.1101/861898.

Webb, P.W. & Cotel, A.J. (2011) Stability and turbulence. In: *Encyclopedia of Fish Physiology: From Genome to Environment*. Academic Press, Dan Diego, USA, volumes 1–3. pp. 581–586. DOI: 10.1016/B978-0-12-374553-8.00221-5.

Wei, T., Mark, R. & Hutchison, S. (2014) The fluid dynamics of competitive swimming. *Annual Review of Fluid Mechanics*, 46, 547–565. DOI: 10.1146/annurev-fluid-011212-140658.

Xie, Q. (1998) *Turbulent Flows in Non-Uniform Open Channels: Experimental Measurements and Numerical Modelling*. Ph.D. Thesis, School of Civil Engineering, The University of Queensland, Brisbane, Australia. 358 pages.

Yu, B., Seed, A., Trevithick, R. & Morris, K. (2007) Flood forecasting system for urban catchments. *Australian Journal of Water Resources*, 11(2), 161–168.

Zhang, G. & Chanson, H. (2018) Three-dimensional numerical simulations of smooth, asymmetrically roughened, and baffled culverts for upstream passage of small-bodied fish. *River Research and Applications*, 34(8), 957–964. DOI: 10.1002/rra.3346.

Bibliography

Additional bibliographic references are provided to the reader. Many are seminal works, covering basic fluid mechanics and open channel flows, as well as experimental biology and ecological sciences.

Abdelaziz, S.M., Jhanwar, R., Bui, M.D. & Rutschmann, P. (2011) A numerical model for fish movement through culverts. *Proc. 34th IAHR World Congress*, 26 June–1 July, Engineers Australia Publication, Brisbane, Australia, Valentine, E., Apelt, C., Ball, J., Chanson, H., Cox, R., Ettema, R., Kuczera, G., Lambert, M., Melville, B. & Sargison, J. (eds.). pp. 2736–2743.

Adhikari, D. & Longmire, E.K. (2013) Infrared tomographic PIV and 3D motion tracking system applied to aquatic predator-prey interaction. *Measurement Science and Technology*, 24(2), Paper 024011, 17 pages. DOI: 10.1088/0957-0233/24/2/024011.

Adhikari, D., Gemmell, B.J., Hallberg, M.P., Longmire, E.K. & Buskey, E.J. (2015) Simultaneous measurement of 3D zooplankton trajectories and surrounding fluid velocity field in complex flows. *Journal of Experimental Biology*, 218(22), 3534–3540.

Alexander, R.M. (2003) *Principles of Animal Locomotion*. Princeton University Press, Princeton, NJ, USA. 371 pages.

ARR (1987) *Australian Rainfall and Runoff: A Guide to Flood Estimation*, Pilgrim, D.H. & Caterford, R.P. (eds.). Institution of Engineers Australia, Barton ACT, Australia, 2 volumes.

Baudoin, J.M., Burgun, V., Chanseau, M., Larinier, M., Ovidio, M., Sremski, W., Steinbach, P. & Voegtle, B. (2014) *Evaluer le franchissement des obstacles par les poissons. Principes et méthodes*. Onema, France. 200 pages (in French).

Belidor, B.F. de (1737–1753) *Architecture Hydraulique* (Hydraulic Architecture). Charles-Antoine Jombert, Paris, France, 4 volumes (in French).

Breton, F., Baki, A.B.M., Link, O., Zhu, D.Z. & Rajaratnam, N. (2013) Flow in nature-like fishway and its relation to fish behaviour. *Canadian Journal of Civil Engineering*, 40, 567–573.

Bushnell, D.M. & McGinley, C.B. (1989) Turbulence control in wall flows. *Annual Review of Fluid Mechanics*, 21, 1–20.

Cabonce, J., Fernando, R., Wang, H. & Chanson, H. (2017) Culvert baffles to facilitate upstream fish passage. *Proceedings of 13th Hydraulics in Water Engineering Conference HIWE2017*, 13–16 November, Engineers Australia, Sydney. 9 pages (ISBN 978-1-925627-03-9).

Cabonce, J., Wang, H. & Chanson, H. (2018) Smart baffles to assist upstream culvert passage of small-bodied fish. *Proc. 7th IAHR International Symposium on Hydraulic Structures ISHS2018*, 15–18 May, Aachen, Germany. 11 pages.

Castro-Orgaz, O. & Chanson, H. (2016) Minimum specific energy and transcritical flow in unsteady open-channel flow. *Journal of Irrigation and Drainage Engineering*, ASCE, 142(1), Paper 04015030, 12 pages. DOI: 10.1061/(ASCE)IR.1943-4774.0000926 – 2018 American Society of Civil Engineers, Environmental and Water Resources Institute (ASCE-EWRI) Honorable Mention Paper Award.

Chanson, H. (2000) Introducing originality and innovation in engineering teaching: The hydraulic design of culverts. *European Journal of Engineering Education*, 25(4), 377–391.

Chanson, H. (2002) Hydraulics of a large culvert beneath the Roman aqueduct of Nîmes. *Journal of Irrigation and Drainage Engineering*, ASCE, 128(5), 326–330. DOI: 10.1061/(ASCE) 0733-9437(2002)128.

Chanson, H. (2012) Momentum considerations in hydraulic jumps and bores. *Journal of Irrigation and Drainage Engineering*, ASCE, 138(4), 382–385. DOI: 10.1061/(ASCE)IR.1943-4774.0000409.

Chanson, H. & Felder, S. (2017) Hydraulics of selected hydraulics structures. In: Radecki-Pawlik, A., Pagliara, S. & Hradecky, J. (eds.) *Open Channel Hydraulics, River Hydraulic Structures and Fluvial Geomorphology: For Engineers, Geomorphologists and Physical Geographers*. CRC Press, Balkema, London, Chapter 2. pp. 25–46 (ISBN 9781498730822).

Chow, V.T., Maidment, D.R. & Mays, L.W. (1988) *Applied Hydrology*. McGraw-Hill, USA. 572 pages.

Cotel, A.J., Webb, P.W. & Tritico, H. (2006) Do brown trout choose locations with reduced turbulence? Transactions of the American Fisheries Society, 135, 610–619.

Cunge, J.A., Holly, F.M., Jr. & Verwey, A. (1980) *Practical Aspects of Computational River Hydraulics*. Pitman, Boston, USA. 420 pages.

Darrozes, S.S. & Monavon, A. (2014) *Analyse Phénoménologique des Ecoulements. Comment traiter un Problème de Mécanique des Fluides avant de résoudre les Equations* (Phenomenological Analysis of Flows: How to Solve a Fluid Mechanics Problem before Solving the Equations). Presses Polytechniques et Universitaires Romandes, Lausanne, Switzerland. 480 pages (in French).

DPI Fisheries (2013) Policy and guidelines for fish habitat conservation and management. NSW Department of Primary Industries. 80 pages.

Esplin, L.D. & Hotchkiss, R.H. (2011) Culvert roughness elements for native utah fish passage: Phase I. Report No. UT-11.02, Utah Department of Transportation-Research Division, USA. 50 pages.

Fernandez-Prats, R., Raspa, V., Thiria, B., Hera-Huarte, F. & Godoy-Diana, R. (2015) Large-amplitude undulatory swimming near a wall. *Bioinspiration & Biomimetics*, 10, Paper 016003, 15 pages. DOI: 10.1088/1748-3190/10/1/016003.

Fish, F.E. & Lauder, G.V. (2013) Not just going with the flow. *American Scientist*, 101(2), 114–123.

Floryan, D., Van Buren, T., Rowley, C.W. & Smits, A.J. (2017) Scaling the propulsive performance of heaving and pitching foils. *Journal of Fluid Mechanics*, 822, 386–397.

Gemmell, B.J., Adhikari, D. & Longmire, E.K. (2014) Volumetric quantification of fluid flow reveals fish's use of hydrodynamic stealth to capture evasive prey. *Journal of the Royal Society Interface*, 11(90), Paper 20130880, 9 pages. DOI: 10.1098/rsif.2013.0880.

Godoy-Diana, R. & Thiria, B. (2018) On the diverse roles of fluid dynamic drag in animal swimming and flying. *Journal of The Royal Society Interface*, 15, Paper 20170715, 13 pages. DOI: 10.1098/ rsif.2017.0715.

Hamill, L. (1999) *Bridge Hydraulics*. E&FN Spon, London, UK. 367 pages.

Hinze, J.O. (1975) *Turbulence*. McGraw-Hill Publishers, New York, USA, 2nd edition.

Laurenson, E.M. (1987) Back to basics on flood frequency analysis. *Civil Engineering Transactions*, IEAust., CE29, 47–53.

Leng, X. & Chanson, H. (2019) Hybrid modelling of low velocity zones in box culverts to assist upstream fish passage. *Environmental Fluid Mechanics*, Springer, 18 pages. DOI: 10.1007/s10652-019-09700-1. (In Print).

Liao, J.C. (2007) A review of fish swimming mechanics and behaviour in altered flows. Philosophical Transactions of the Royal Society B, 362, 1973–1993. DOI: 10.108/rstb.2007.2082.

Marsden, T., Thorncraft, G. & Woods, K. (2003) Reconstruction of culverts and causeway to assist migration of adult and juvenile fish project. Department of Primary Industries, Queensland Government, NHT Project No. 2012102, Final Project Report. 34 pages.

Marsden, T., Stewart, R., Woods, K., Jennings, D., Shantele, I. & Thorncraft, G. (2006) *Freshwater Fish Habitat Rehabilitation in the Mackay Whitsunday Region*. Department of Primary Industries, Queensland Government, Information Series QO 003012, Queensland, Australia. 135 pages.

Melville, B.W. & Coleman, S.E. (2000) *Bridge Scour*. Water Resources Publications, Highlands Ranch, USA. 550 pages.

Michalec, F.G., Souissi, S. & Holzner, M. (2015) Turbulence triggers vigorous swimming but hinders motion strategy in planktonic copepods. *Journal of the Royal Society Interface*, 12, Paper 20150158, 10 pages. DOI: 10.1098/rsif.2015.0158.

Monk, S.K., Wait, L.E., Hotchkiss, R.H., Billman, E., Belk, M. & Stuhft, D. (2012) Culvert roughness elements for native utah fish passage. *Proc. World Environmental and Water Resources Congress*, 20–24 May, ASCE, Albuquerque, NM, USA. pp. 1301–1307.

Moore, M., McCann, J. & Power, T. (2018) Greater Brisbane fish barrier prioritisation. Catchment Solutions Report. 97 pages.

Morrison, R.R., Hotchkiss, R.H., Stone, M., Thurmand, D. & Horner-Devinee, A.R. (2009) Turbulence characteristics of flow in a spiral corrugated culvert fitted with baffles and implications for fish passage. *Ecological Engineering*, 35, 381–392.

Nathan, R. (2012) *Brisbane: Lessons from Large Floods*. EA Magazine, Engineers, Canberra ACT, Australia, December. pp. 44–45.

O'Connor, C. (1993) *Roman Bridges*. Cambridge University Press, Cambridge, UK. 235 pages.

Piquet, J. (1999) *Turbulent Flows: Models and Physics*. Springer, Berlin, Germany. 761 pages.

Potter, D.J. & Pilgrim, D.H. (1971) Flood estimation using a regional flood frequency approach. Final Report, Vol. 2, Report on Analysis Components. Australian Water Resources Council Research Project 68/1, Hydrology of Small Rural Catchments. Snowy Mountains Engineering Corporation, Australia.

Prandtl, L. (1952) *Essentials of Fluid Dynamics with Applications to Hydraulics, Aeronautics, Meteorology and Other Subjects*. Blackie & Son, London, UK. 452 pages.

Preissmann, A. & Cunge, J.A. (1967) Low-amplitude undulating hydraulic jump in trapezoidal. *Journal of Hydraulic Research*, IAHR, 5(4), 263–279.

Quadrio, J. (2007) Passage of fish through drainage structures. *Queensland Roads*, September, 6–17.

Roache, P.J. (2009) Perspective: Validation-what does it mean? *Journal of Fluids Engineering*, ASME, March, 131, Paper 034503, 4 pages.

Rodi, W. (2017) Turbulence modeling and simulation in hydraulics: A historical review. *Journal of Hydraulic Engineering*, ASCE, 143(5), Paper 03117001, 20 pages. DOI: 10.1061/(ASCE)HY.1943-7900.0001288.

Rohr, J.J. & Fish, F.E. (2004) Strouhal numbers and optimization of swimming by odontocete cetaceans. *Journal of Experimental Biology*, 207(10), 1633–1642.

Sailema, C., Freire, R., Chanson, H. & Zhang, G. (2019) Modelling small ventilated corner baffles for box culvert barrel. *Environmental Fluid Mechanics*, Springer, 25 pages. DOI: 10.1007/s10652-019-09680-2. (In Print).

Sanchez, P., Leng, X. & Chanson, H. (2019) Hydraulics of an asymmetrical flume with sidewall rib. *Proc. 38th IAHR World Congress*, 1–6 September, IAHR Publication, Panama City, Calvo, L. (ed.). pp. 6160–6170. DOI: 10.3850/38WC092019-0211.

Schlichting, H. & Gersten, K. (2000) *Boundary Layer Theory*. Springer Verlag, Berlin, Germany, 8th edition. 707 pages.

Shaw, E.M., Beven, K.J., Chappell, N.A. & Lamb, R. (2011) *Hydrology in Practice*. CRC Press, Oxon, UK, 4th edition. 543 pages.

Smits, A.J. (2019) Undulatory and oscillatory swimming. *Journal of Fluid Mechanics*, 874(P1), 70 pages. DOI: 10.1017/jfm.2019.284.

Transportation Association of Canada (2004) *Guide to Bridge Hydraulics*. Thomas Telford, London, UK, 2nd edition. 181 pages.

Wallace, J.M. (2009) Twenty years of experimental and direct numerical simulation access to the velocity gradient tensor: What have we learned about turbulence? *Physics of Fluids*, 21, Paper 021301, 17 pages. DOI: 10.1063/1.3046290.

Wallace, J.M. (2013) Highlights from 50 years of turbulent boundary layer research. *Journal of Turbulence*, 13(53), 1–70.

Wang, H. & Chanson, H. (2017a) Baffle systems to facilitate upstream fish passage in standard box culverts: How about fish-turbulence interplay? *Proceedings of 37th IAHR World Congress*, 13–18 August, IAHR & USAINS Holding Sdn. Bhd. Publ., Kuala Lumpur, Malaysia, Ghani, A.A., Chan, N.W., Ariffin, J., Wahab, A.K.A., Harun, S., Kassim, A.H.M. & Karim, O. (eds.), Vol. 3, Theme 3.1. pp. 2586–2595.

Wang, H. & Chanson, H. (2017b) How a better understanding of fish-hydrodynamics interactions might enhance upstream fish passage in culverts. Research Report No. CE162, School of Civil Engineering, The University of Queensland, Brisbane, Australia. 43 pages (ISBN 978-1-74272-192-7).

Yasuda, Y. (2011) *Guideline for Fish Passages for Engineers: Based on Flow Conditions and Structure of Fish Passages*. Corona Publishing, NPO Society for Fishway in Hokkaido, Tokyo, Japan. 144 pages.

Zhang, G. & Chanson, H. (2018) Numerical investigations of box culvert hydrodynamics with smooth, unequally roughened and baffled barrels to enhance upstream fish passage. Hydraulic Model Report No. CH111/18, School of Civil Engineering, The University of Queensland, Brisbane, Australia. 129 pages (ISBN 978-1-74272-197-2).

Internet bibliography

Upstream fish passage in box culverts: how do fish and turbulence interplay?	http://staff.civil.uq.edu.au/h.chanson/fish_culvert.html
Australian rainfall and runoff (ARR) 2016	http://arr.ga.gov.au/arr-guideline

YouTube video movie

Fish-friendly waterways and culverts – integration of hydrodynamics and fish turbulence interplay and interaction	https://youtu.be/GGWTWDOmoSQ

Open access repositories

OAIster	www.oclc.org/en/oaister.html
UQeSpace	http://espace.library.uq.edu.au/

Author index

Note: Page numbers in *italics* indicate a figure and page numbers in **bold** indicate a table on the corresponding page. Numbers followed by "n" indicate an endnote.

Subject index

Note: Page numbers in *italics* indicate a figure and page numbers in **bold** indicate a table on the corresponding page. Numbers followed by "n" indicate an endnote.

Printed and bound by CPI Group (UK) Ltd, Croydon, CR0 4YY

24/10/2024

01778288-0004